Advances in
Disease Vector Research

Advances in Disease Vector Research

Edited by

Kerry F. Harris
Virus Vector Laboratory, Department of Entomology, Texas A&M
University, College Station, Texas 77843, USA

Advances in
Disease Vector Research

Volume 7

Edited by Kerry F. Harris

With Contributions by
C.J. Andrews R.H. Bagnall N. Carter C. Chastel
J.R. DeLoach K.S. Gibb R. Harrington
I. Humphery-Smith I. Maudlin J.W. Randles
R.C. Sinha G. Spates

With 31 Illustrations

Springer-Verlag
New York Berlin Heidelberg London
Paris Tokyo Hong Kong Barcelona

Kerry F. Harris
Virus Vector Laboratory
Department of Entomology
Texas A&M University
College Station, Texas 77843, USA

Volumes 1 and 2 of *Current Topics in Vector Research* were published by Praeger Publishers, New York, New York.

ISSN: 0934-6112

Typeset by Best-set Typesetter Ltd., Quarry Bay, Hong Kong.

9 8 7 6 5 4 3 2 1

ISBN-13:978-1-4613-9046-6   e-ISBN-13:978-1-4613-9044-2
DOI:10.1007/978-1-4613-9044-2

Preface

We open Volume 7 with a series of four chapters on plant virus transmission by insects. In Chapter 1, Karen Gibb and John Randles present preliminary information about an association between the plant bug *Cyrtopeltis nicotianae* (Heteroptera: Miridae) and velvet tobacco mottle virus (VTMoV): the only reported instance of mirid transmission of a known virus. Mirids could be considered as likely vectors of plant viruses because they are phytophagous, possess a piercing–sucking–feeding apparatus, have winged adults, and are cosmopolitan pests of a wide range of crops. Surprisingly, however, there are only three plant viruses purportedly transmitted by heteropterous vectors, compared with the nearly 250 by homopterous ones. To what extent these figures reflect actual differences in the abilities of members of the two suborders to transmit plant pathogens remains to be determined. Compared with the Homoptera, the Heteroptera have been ignored by researchers as potential vectors of plant viruses.

The authors are quick to point out that additional studies are needed before generalizations can be made about virus–mirid–plant interactions and that virus transmission by mirids is not easily characterized using the conventional transmission criteria and terminology established for such homopterous vectors as aphids and leafhoppers. Transmission of VTMoV by *C. nicotianae* appears to have characteristics in common with both nonpersistent noncirculative and circulative (persistent) transmission. However, because there is no evidence to associate the virus with the salivary glands of its vector, the authors hypothesize that the virus "translocates" in its mirid vector rather than circulates (i.e., virus enters the vector through the maxillary food canal and exits via the salivary canal). The possibility of transmission via an ingestion–defecation pathway is discussed, as are similarities between virus transmission by mirids and that by beetle and mite vectors.

In Chapter 2, Nick Carter and Richard Harrington discuss the many factors affecting aphid population dynamics and behavior and their subsequent influence on aphid-borne virus spread: all from an entomologist's viewpoint. The authors first examine the influence of various weather factors on aphid overwintering survival, spring population development, dispersal from winter host plants, and virus spread within crops. They then describe how intrinsic characteristics of the vectors influence virus spread, such as their tendency to produce winged or sexual forms. The roles of aphid natural enemies and noncrop plants in virus spread are also discussed, the latter of which harbor crop viruses, vectors of crop viruses, or both. In the final section, the authors address the question of how such crop management practices as rotation, sowing dates, and pesticide usage influence virus spread though their effects on vector populations. The influences of these various factors on nonpersistent, semipersistent, and persistent virus transmission are discussed in a comparative manner. Applications of the resulting information to forecasting are considered also, although the authors agree that most efforts in this direction are relatively undeveloped due to the complexity of both the factors involved and their interactions.

In Chapter 3, Richard Bagnall will arouse no doubt the interest of more than a few epidemiologists and climatologists with his theory regarding the cyclic nature of aphid-borne potato virus epidemics in northern latitudes during recent times. There have been three major epidemics of the potato leaf roll virus (PLRV) in the seed-growing areas of northern Maine and New Brunswick, as well as in northwestern Europe during the 20th century. These events have occurred at approximate 32-year intervals, virtually in phase sequence with earlier outbreaks of "the curl," a "degeneration" disease of potatoes known in Europe during the 18th and 19th centuries. The characteristics of the epidemics, their regular occurrence, and coincidences with periodic natural phenomena—periods of warm climate and drought, and certain sunspot minima—point to the workings of a widespread climatic cycle.

In New Brunswick, the potato mosaic diseases, due to potato viruses Y and A (PVY and PVA), were found also to fluctuate in a cyclic pattern, but a different one from that of PLRV. This mosaic cycle, averaging 9.3 years, runs clearly out of phase with the sunspot cycle, but has been in phase during the past 66 years with a population cycle in Canadian wildlife known as the *Canada Hare–Lynx Cycle*. A possible connection is that the hare thrives on an abundance of many of the same shrubs and trees on which aphids lay eggs.

The difference between the two cycles derives from the different modes of transmission of the viruses. The persistently transmitted PLRV is transmitted largely by the potato-colonizing aphid, *Myzus persicae* (Sulz.), which does not overwinter readily in New Burnswick. And PLRV outbreaks have occurred during discrete 5–10 year periods of milder climate. By contrast, the nonpersistent PVY can be transmitted by numerous aphids that do not necessarily colonize potatoes, but visit in search of feeding sites. Many of these aphids are better adapted to the northern climate. Moreover, the Canada hare–lynx cycle is reputed to be a northern phenomenon.

The mosaic disease cycle in New Brunswick is actually more complex. During the most recent cycle, the relatively intense spread of the major virus, PVY, occurred only during alternate even years (1974, 1976, 1978, 1980, and 1982): a biennial rhythm. After a double low (1983 and 1984), a new cycle has begun with highs in odd years (1985 and 1987). In effect, there has been an intense spread in only 7 of the past 16 years, with no 2 in succession. This rhythm has a significant bearing on the status of New Brunswick as a seed-growing area. In addition, spread of the latent potato virus S was found to be subject to the same biennial rhythm as is PVY. Therefore, it is held to be spread in a nonpersistent manner by many of the same aphids.

The author also points out that both the PLRV and mosaic cycles may be affected further by longer-term climatic change. If the sharp rise in northern hemisphere temperature since 1970 continues, the PLRV epidemics may become more frequent or prolonged. Conversely, the warming and drying that occurred in the 1930s and 1940s coincided with a decline both in the lynx harvest and in the incidence of mosaic diseases.

Chapter 4 is dedicated to the memory of Dr. Yogesh Paliwal, who passed away on June 23, 1988. Authors Chris Andrews and Ramesh Sinha describe the aphid vectors of barley yellow dwarf virus (BYDV), which causes a major disease in cereal crops throughout the world, and discuss the effects of the virus on the ability of winter cereal plants to withstand the winter stresses that influence the plants in the northern part of their range. Current views of the relationships between the various virus strains are given, with a description of their aphid vectors, and a survey of the epidemiology of the disease. After a brief review of the effects of freezing stresses on plants, the effects of BYDV on overwintering cereals are described from a series of field studies. Experiments in controlled environments are reviewed that separate BYDV effects on cold-hardiness, ice-tolerance and low-temperature-flooding tolerance. Another section describes some of the biochemical associations between virus infection and stress-tolerance reduction, and the chapter concludes with speculation on the precise nature of the interaction and possibilities for increasing the tolerance of plants to the virus.

Beginning with Chapter 5, we switch our attention to human and animal disease associations with blood-feeding flies. In this chapter, authors John Deloach and George Spates give us a brief but thorough review of the limited literature on artificial diets, "truly defined diets," for stable flies and tsetse flies. The rather unusual combination of these strikingly different arthropods in one review stems from the fact that stable flies were used for dietary model studies in Texas, the authors' home base, whereas tsetse flies were used elsewhere in Europe, Africa, and Canada.

The authors begin by defining and categorizing diets for blood-feeding arthropods and the approaches taken to developing them. Most of the research on artificial diets for blood-feeding insects has been directed by only a few laboratories, and the Tsetse Research Laboratory in Bristol, England, is one such center that has made significant contributions in this regard. Indeed, the authors were aided in their review because of their collaboration over the past 10 years with researchers at both the Tsetse Research Laboratory and the International Atomic Energy Agency in Vienna, Austria. Basically, animal blood was dissected into its various components (blood cells and plasma) and subsequently subdivided into proteins and fats. Albumin, hemoglobin, phospholipids, and cholesterol are dietary requirements for both *Stomoxys* and *Glossina*. It is unlikely that amino acids comprising the proteins can be substituted for albumin or hemoglobin because of the high osmotic pressure of the amino acid solutions. Although significant progress has been made, there remain several unanswered questions regarding artificial diets for blood-feeding insects.

In Chapter 6, Ian Maudlin discusses interactions among the tsetse immune system, symbionts, and parasites that are crucial to transmission of African trypanosomiasis. Until recently, transmission of the African trypanosomiases was thought to depend largely on fly number: every tsetse in a population was regarded as a potential vector. Maudlin here reviews the evidence that led to this view and describes recent experimental work which, together with field data, shows that most tsetse flies in a population, in fact, are refractory to infection. The basis of this refractoriness is shown to lie in the tsetse immune system that is equipped specifically with lectins to deal with incursions by trypanosomes. Refractoriness to trypanosomes can be inhibited by symbionts that inhibit lectin output, rendering flies carrying these bacteria susceptibles. Intriguingly, this insect defense system has been put to use by the trypanosome to act as a trigger to complete its life cycle in the fly. That the tsetse immune system should be involved so closely in the transmission of the disease is an important finding and has clear implications for research on other vector–parasitc relationships.

The volume ends with Chapter 7, a thorough and timely treatise on mosquito spiroplasmas by Claude Chastel and Ian Humphery-Smith. For a group of organisms as well studied as mosquito vectors, it is surprising that mosquito spiroplasmas (msp) were not discovered prior to 1982. This event closely followed the development of suitable culture media and since has made possible the isolation of several msp and the description of *Spiroplasma culicicola* from the United States, *S. sabaudiense* from France and *S. taiwanense* from the Far East. Less than 10 years has elapsed since the initial discovery of msp by Slaff and Chen, and although our knowledge of these organisms has increased greatly, the amount of information represents merely the tip of an iceberg. Some researchers have even suggested that spiroplasmas may one day be shown to be as diverse as their arthropod hosts. Once more research groups start looking for msp in a systematic fashion, undoubtedly more msp will be uncovered. And the authors have purposefully included details pertaining to isolation techniques to further that end.

At the very least, the *Cantharis* spiroplasma complex, isolated from mosquitoes, plants, and other anthropods, probably circulates in nature via flower nectar. However, much needs to be clarified before the ecology of msp is appreciated fully. Real hope is sustained for the eventual use of msp as biological control agents against mosquito vectors of tropical disease and mosquito species of nuisance value for urban dwellers of the world's major cities.

Finally, the possible role of thermophilic spiroplasmas (as yet no msp) as causative agents of degenerative central nervous system disorders makes this field both exciting and of potential relevance to both human and animal health.

I thank the authors for their scholarly contributions as well as their patience and diligence in working with me to bring Volume 7 to such a successful conclusion. This volume contains information of interest to most biologists, but in particular to plant pathologists, virologists, microbiologists, vector ecologists, disease epidemiologists, both medical and veterinary arthropodologists and parasitologists, and even climatologists. I humbly and gratefully acknowledge the encouragement and technical assistance of the members of the Editorial Board and staff of Springer-Verlag. Without their assistance, my job would not be nearly so rewarding.

Kerry F. Harris

Contents

Contributors

Christopher J. Andrews
Plant Research Centre, Agriculture Canada, Central Experimental Farm, Ottawa, Ontario, Canada, K1A OC6

Richard H. Bagnall
Agriculture Canada Research Station, P. O. Box 20280, Fredericton, New Brunswick, Canada, E3B 4N6

Nick Carter
Department of Entomology and Nematology, AFRC Institute of Arable Crops Research, Rothamsted Experimental Station, Harpenden, Hertfordshire, AL5 2JQ, England, U.K.

Claude Chastel
Department de Microbiologie et Santé Publique, Faculté de Médecine, 29285 Brest, France

John R. DeLoach
U.S. Department of Agriculture, Agricultural Research Service, Veterinary Toxicology and Entomology Research Laboratory, College Station, Texas 77840, U.S.A.

Karen S. Gibb
Faculty of Science, Northern Territory University, P. O. Box 40146, Casuarina, Northern Territory 0811, Australia

Richard Harrington
Department of Entomology and Nematology, AFRC Institute of Arable Crops Research, Rothamsted Experimental Station, Harpenden, Hertfordshire, AL5 2JQ, England, U.K.

Ian Humphery-Smith
Department de Microbiologie et Santé Publique, Faculty é de Médecine, 29285 Brest, France

Ian Maudlin
Tsetse Research Laboratory, ODA/University of Bristol, Langford House, Langford, Bristol BS18 7DU, England, U.K.

John W. Randles
Department of Plant Pathology, Waite Agricultural Research Institute, Glen Osmond, South Australia, 5064, Australia

Ramesh C. Sinha
Plant Research Centre, Agriculture Canada, Central Experimental Farm, Ottawa, Ontario, Canada, K1A OC6

George Spates
U.S. Department of Agriculture, Agricultural Research Service, Veterinary Toxicology and Entomology Research Laboratory, College Station, Texas 77840, U.S.A.

Contents for Previous Volumes

Volume 2

Volume 3

Volume 6

1
Transmission of Velvet Tobacco Mottle Virus and Related Viruses by the Mirid *Cyrtopeltis nicotianae*

Karen S. Gibb and John W. Randles

Introduction

Mirids belong to the family Miridae, which is one of the largest families in the suborder Heteroptera (28). Their mouthparts, which are modified for sucking plant sap, consist of two pairs of flexible stylets enclosed by a labium (Fig. 1.1). This is a characteristic of all Hemiptera including aphids, the largest group of insect vectors. Adult mirids are winged and highly mobile and they are important pests of a wide range of crops (5, 7, 19, 27, 20, 6, 16). Their feeding behavior, mobility, and wide host range might be expected to place them among the important vectors of plant diseases. However, there have only been three reports in the literature of disease transmission by mirids and only one of these involves a plant virus. This paucity of information may reflect either the fact that mirids are not significant vectors of plant diseases, or it may simply be that few workers have studied mirid–plant-pathogen interactions. If future surveys indicate that the former is true, then it would be of interest to know why mirids are not important vectors in the field.

The first example of mirid transmission was reported more than 70 years ago, for the uncharacterized spinach blight disease by the tarnished plant bug *Lygus pratensis* (17). This transmission was suspected of occurring via a mechanical mechanism (14). The second report was of the transmission of fire blight bacteria (*Erwinia amylovora*) to pear fruit by *Lygus* spp. This was considered to occur by mechanical inoculation also (25). The only report of the natural transmission of a virus by mirids is that of velvet tobacco mottle virus (VTMoV) by the mirid *Cyrtopeltis nicotianae* (24, 8).

Karen S. Gibb, Faculty of Science, Northern Territory University, Casuarina, Northern Territory, 0811, Australia.
John W. Randles, Department of Plant Pathology, Waite Agricultural Research Institute, Glen Osmond, South Australia, 5064, Australia.

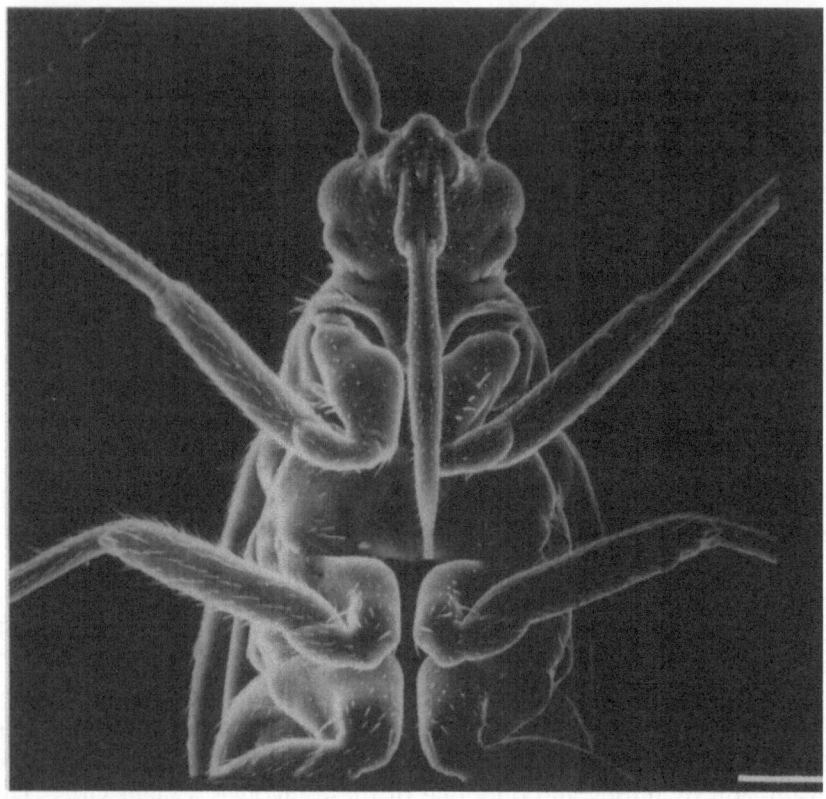

FIGURE 1.1. Mouthparts of *C. nicotianae*. Scanning electron micrograph of a 5th-stage nymph showing the labium that encloses the stylet. Bar represents 200 μm.

The characteristics of transmission of VTMoV by *C. nicotianae* have been studied in detail by the authors of this chapter, and used here as the sole example of mirid transmission. Clearly, until more studies are done, generalizations cannot be made about mirid transmission as can be done for aphid transmission, for example, but it is hoped that this chapter will foster interest in the further study of associations between mirids and plant viruses.

Most studies of insect transmission have been done on the three major homopteran vectors, that is, aphids, leafhoppers, and planthoppers, and the definitions used to categorize virus–vector associations (11), have been adapted from these studies. The categories of transmission used in this chapter follow those proposed by Harris (13) in which virus–vector associations are classified as either noncirculative (including both nonpersistent and semipersistent associations) or circulative (synonymous with persistent). Circulative viruses may be categorized further as either propagative or nonpropagative. Problems arise when trying to characterize mirid transmission using these conventional criteria and this chapter discusses these problems and suggests alternative mechanisms to explain our observations of mirid transmission. Virus transmission by beetles and mites has some characteristics in common with transmission by mirids and this is discussed also.

Characteristics of Transmission

Transmission of Related Viruses

Velvet tobacco mottle virus (VTMoV) was isolated from diseased velvet tobacco, *Nicotiana velutina* (Wheeler) collected in northeastern South Australia (24). Adults and nymphs of *Cyrtopeltis nicotianae* (Konings) collected from *N. velutina* in the same area transmitted VTMoV (24). VTMoV is the first of four viruses isolated in Australia shown to contain an unusual circular satellite RNA (24). The other three viruses are solanum nodiflorum mottle virus (SNMV), lucerne transient streak virus (LTSV) and subterranean clover mottle virus (SCMoV). All four viruses are members of the sobemovirus group (23, 12, 3, 4) and could be considered as a satellite-bearing subgroup of the sobemoviruses.

Populations of *C. nicotianae* used in transmission experiments were reared on *N. clevelandii* A. Gray. Plants of different ages were maintained in an insectary at $25 \pm 4°C$, and the colony was managed by replacing plants of at least 2 months of age with young seedlings on a long rotation program. The five nymphal stages and winged adult are shown in Figure 1.2. The ability of mirids to transmit the four sobemoviruses mentioned above was tested, as was the transmission of the virus southern bean mosaic virus (SBMV) and another member, sowbane mosaic virus (SoMV). Besides VTMoV, mirids could transmit the serologically related SNMV as well as SBMV and SoMV, but not, however, SCMoV or LTSV (8).

Acquisition

Feeding trials were done to determine the minimum time required for mirids to acquire the virus. The acquisition threshold was less than 1 min and the rate of transmission increased with increasing acquisition time up to 30 min. After this time the rate of transmission increased only slightly (Fig. 1.3).

Inoculation

The inoculation threshold and the relationship between acquisition time and rate of inoculation were studied. Three preinoculation conditions were

FIGURE 1.2. The five nymphal stages and an adult of *C. nicotianae*. Nymphal stages are classified according to wing-bud development and color, the first two stages are yellow, the remainder are green. The body of the adult is green with light-brown antennae, some segments of which are black. The wings are transparent with black patches. Bar represents 1 mm. From Ref. 8.

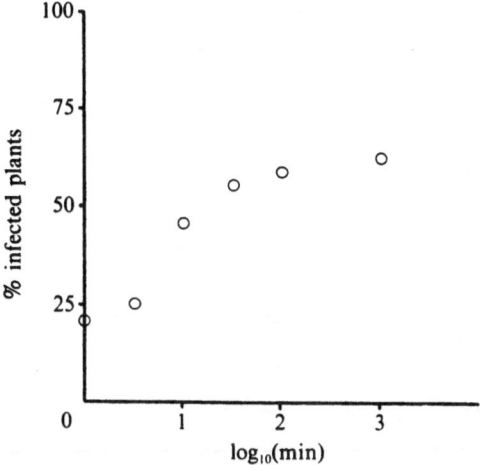

FIGURE 1.3. Relationship between acquisition period ($\log_{10}$ min) and percentage of plants inoculated. Each point represents the percentage of plants infected by mirids at each acquisition time. From Ref. 8.

tested (1) 1-h access to VTMoV; (2) 2-day access to VTMoV followed by starvation for 16 h; or (3) 2-day access to VTMoV. The results are shown in Figure 1.4 where the symbols indicate the actual points and the solid line indicates the fitted values. When mirids had access to VTMoV or 2 days with no postacquisition fast, the minimum inoculation period was between 1 and 2 h (Graph 3, Fig. 1.4). However, when mirids fasted after acquisition (Graph 2), the minimum inoculation period was between 2 and 4 h and when the acquisition period was 1 h only, the minimum inoculation period was between 4 and 8 h (Graph 1, Fig. 1.4). The slopes of the lines from Graphs 1 and 2 were not significantly different from each other, but both were different from that of Graph 3. Thus shorter acquisition periods and postacquisition fasts had a similar effect of reducing the rate of inoculation and increasing the time required to inoculate plants when compared to mirids that had longer acquisition periods (8).

Persistence

To test persistence, mirids were given an acquisition period of 24 h and then were transferred daily to uninfected test plants for 12 days or until the death of the mirid. Infectivity was retained for up to 10 days (Fig. 1.5), although during this period virus was transmitted only intermittently and rarely on successive days. There was some evidence from this experiment that mirids still could transmit after undergoing a postacquisition molt and further experiments were done to study transstadial transmission.

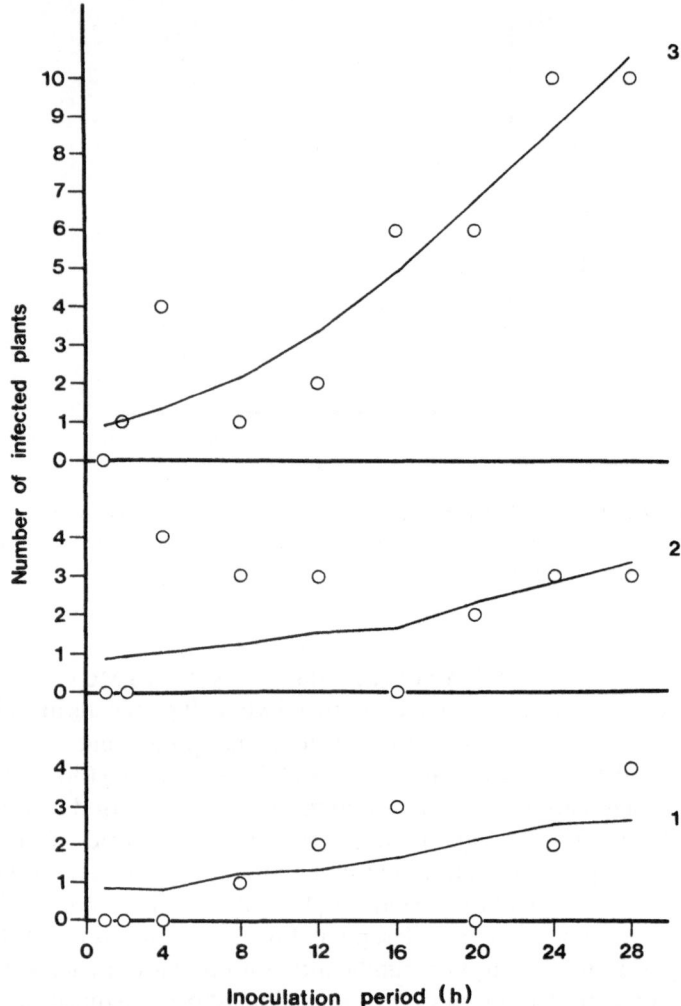

FIGURE 1.4. Relationship between inoculation period and the rate of transmission, following three preinoculation treatments. In Treatment 1 (Graph 1), mirids were allowed an acquisition access period of 1 h. In Treatment 2 (Graph 2), mirids were allowed an acquisition time of 2 days with a 16-h postacquisition fast. In Treatment 3 (Graph 3), the acquisition time was 2 days. The open circles represent experimental values and the solid line represents a fitted relationship. Each point represents a test with 15 insects, and data were fitted following a logit transformation. From Ref. 8.

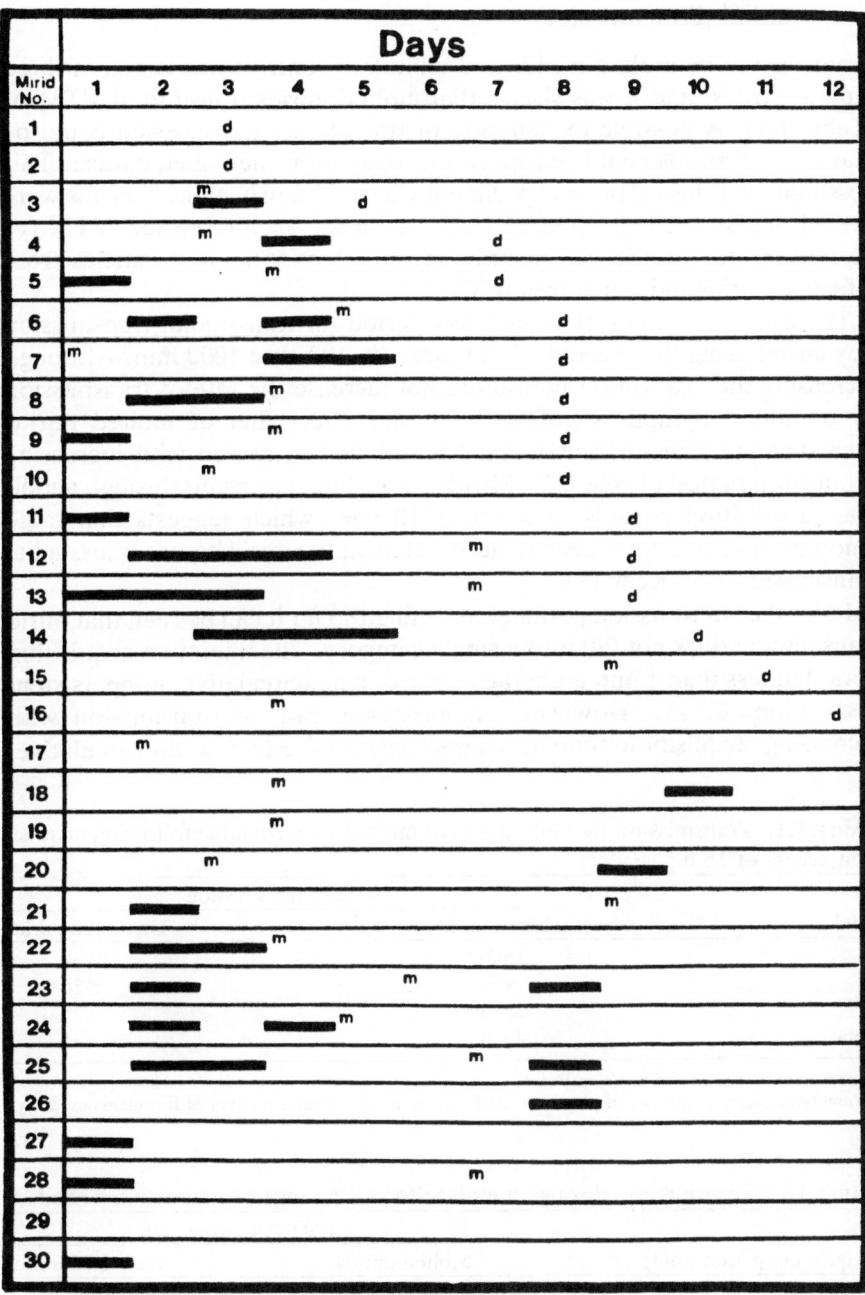

FIGURE 1.5. Time course of transmission of VTMoV by single mirids following a 24-h acquisition feed, showing intermittent sequential transmission and frequent transstadial transmission. Transfer of mirids to test plants continued daily for 12 days or until death of the mirid (d). The time of molting (m) and the days on which plants were infected (■) are shown. From Ref. 8.

Transstadial Transmission

Transmission by molted mirids and unmolted controls was tested in three separate trials, and it was shown that molted mirids transmitted VTMoV (Table 1.1). A possible explanation of transstadial transmission could be that molted mirids could reacquire the virus from their shed cuticle. This possibility was tested in an experiment where 32 newly molted mirids were denied access to their shed cuticle. Of those, 12 transmitted VTMoV. Moreover, nonviruliferous nymphs given access to molted cuticles from infective mirids did not transmit VTMoV (8).

The effect of length of acquisition period on transstadial transmission was tested using three acquisition times; 10, 100, and 1000 min. Although increasing the acquisition period did not increase the rate of transmission by unmolted nymphs significantly, it did affect that of molted mirids such that the rate of transstadial transmission increased with increasing acquisition period (Table 1.2). Mirids were able to transmit through a molt after acquisition periods as short as 10 min, which suggests that large amounts of virus do not need to accumulate in the mirid before transstadial transmission can occur (8).

From the transmission parameters defined so far it can be seen that mirid transmission does not fall into a single category. The minimum acquisition period of less than 1 min is characteristic of a noncirculative, nonpersistent association (8, 18). However, an increasing rate of transmission with increasing acquisition time is characteristic of either a noncirculative,

TABLE 1.1. Transmission by molted and unmolted *C. nicotianae* following acquisition access of 16 h.*

	Rate of transmission	
Trial	Molted mirids	Unmolted mirids
1	16/31[a]	10/30
2	9/24	8/26
3	15/29	6/25
Total	40/84	24/81

* Inoculation access was 48 h. From Ref. 8.
[a] Numerator indicates number of plants infected; denominator indicates number of test plants used.

TABLE 1.2. Transmission through a molt following a range of acquisition periods.*

	Rate of transmission	
Acquisition period (min)	Molted mirids	Unmolted mirids
10	4/40[a] (0.1)[b]	5/20 (0.25)
100	5/33 (0.15)	6/20 (0.30)
1000	6/19 (0.32)	5/16 (0.31)

* From Ref. 8.
[a] Ratio as in Table 1.1.
[b] Proportion.

semipersistent or a circulative association between a virus and its vector (8, 11). The inoculation period of between 1 and 2 h and persistence for up to 10 days are characteristics of either a noncirculative, semipersistent, or a circulative association (11, 8). The transstadial transmission may be due either to retention of virus in the mirid at sites that are not shed during molting, or by circulation of the virus in the mirid (8). One site of retention not shed during molting is the midgut and the possibility of transmission resulting from virus retained at this site was suggested for beetle transmission (14).

The Paradox of Persistence

Circulativity of Virus in the Mirid

The parameters of transmission defined so far indicate that the transmission of VTMoV by the mirid has characteristics of both a noncirculative (nonpersistent and semipersistent), and circulative association (8). Circulativity was tested further by determining whether mirids could transmit following injection of virus into the body cavity, and by determining where the virus is located throughout the body of the mirid (9).

When 60 nymphs were injected with a 2 mg/ml solution of VTMoV labelled with ^{32}P-orthophosphate, only two mirids transmitted the virus and these received 50 and 63 ng of virus each (9). Although 80% of the mirids injected received < 50 ng of virus, five mirids that received > 50 ng also failed to transmit virus (9). In a second experiment, nymphs were

TABLE 1.3. Time course of loss of virus antigen and infectivity in groups of 10 mirids following acquisition for 48 h.[*]

Days after acquisition completed	Number positive[a] by ELISA	Mean virus content of mirids positive by ELISA (ng ± SE)	Number transmitting
1	10/10	1760 ± 500	8/10[b]
2	9/10	2440 ± 900	1/10
3	3/10	630 ± 160	1/10
4	2/10	1030 ± 30	0/10
5	4/10	720 ± 180	1/10
6	3/10	490 ± 150	1/10
7	2/10	470 ± 320	1/10
8	1/10	330	NT[c]
9	0/10	—	NT
10	0/10	—	NT

[*] From Ref. 9.
[a] The minimum amount of virus that could be detected by ELISA was 12 ng.
[b] Ratio as in Table 1.1.
[c] NT = Not tested.

injected with a 15-fold higher concentration of unlabelled virus solution and of 47 nymphs, 11 transmitted VTMoV (9).

The ability of mirids to transmit VTMoV following injection is characteristic of a circulative association (8). The increase in the rate of transmission when a higher concentration of virus was injected into mirids suggested that the ability to transmit following injection may be dose-dependent (9). On the other hand, injection of larger amounts of virus in a labelled solution did not necessarily confer the ability to transmit (9). On the basis of these results we can conclude that there is some evidence that VTMoV circulates in its mirid vector, but there may not be a direct relationship between the amount of virus injected into a mirid and its ability to transmit.

Absence of Demonstrable Virus Propagation

Feeding trials with mirids showed that virus acquired by feeding was transmitted for 5 to 9 days after acquisition (9). In another experiment, mirids given access to virus were transferred to an immune host *Lycopersicon esculentum*. Each day for 10 days, 20 mirids were removed from the tomatoes, 10 were tested individually by enzyme-linked immuno-sorbent assay (ELISA) and 10 were placed singly on uninfected test plants.

The proportion of mirids with virus detectable by ELISA fell sharply between 2 and 3 days after acquisition, and no mirids contained detectable antigen after 9 or 10 days (Table 1.3). These results combined with those from the feeding trial indicate that although there may be limited evidence for circulation of virus in the mirid, there is no evidence to suggest that VTMoV propagates in its mirid vector (9). In this same experiment, the amount of virus antigen detected by ELISA in individual mirids tested on the same day, varied widely (Table 1.3). This may be a reflection of variation in feeding behavior (9). It is noteworthy that a greater proportion of mirids contained virus antigen detectable by ELISA rather than mirids that transmitted VTMoV (9), and a number of other studies have found similar results with other viruses (26, 2, 1). This observation would agree with a hypothesis that some mirids that do not transmit virus may retain detectable amounts of it in sites from which transmission cannot occur (9).

In another experiment in which mirids were fed a labelled virus solution through a membrane and then measured for radioactivity before being given inoculation access to test plants, the virus transmission threshold was 470 ± 320 ng (9). This is far greater than the 50 ng required for transmission following injection and supports the suggestion that virus acquired by feeding may be retained in sites from which transmission cannot occur (9).

Relationship of Virus Distribution to Transmissibility

Virus Translocation in the Mirid

A horseradish peroxidase (HRP) dot-immunobinding assay (15) was used to detect virus in individual body parts. Virus was detected in the gut, feces, and hemolymph of some mirids that had previous access to a source of virus, and more nymphs contained virus in the gut than in either the hemolymph or the feces (10). Virus was not detected in the salivary glands of mirids using the HRP assay (10). These results showed that although VTMoV circulates in the mirid, there was no evidence that the salivary glands are involved in transmission.

Circulative associations are considered to be those in which virus is associated with the vector's salivary system. Therefore, the observed transstadial transmission, transmission following injection and detection of virus in the mirid's hemolymph, are considered to be evidence for virus translocation in the mirid rather than evidence for circulative transmission (9).

Virus Distribution in the Gut or Hemolymph and Transmissibility

Experiments were done to determine whether there was a correlation between transmission and the presence of virus in the hemolymph or gut. Following acquisition, hemolymph samples were taken and assayed for virus, and the mirids were given access to healthy test plants. Following the inoculation feed, mirids were sacrificed to test their gut for virus. Their ability to transmit was shown to be independent of the detection of virus in hemolymph or gut (Table 1.4).One point to note from this experiment is that although hemolymph was assayed immediately after acquisition, due to the destructive nature of sampling the gut, each mirid had 2 days on a healthy test plant before guts were sampled. It is possible that mirids that transmitted did have virus in their gut, but it had "cleared" before sampling.

TABLE 1.4. Relationship between site of detection of virus and ability to transmit.

Virus detected in:	Number of nymphs	Number of nymphs transmitting
Hemolymph only	4	3
Hemolymph and gut	7	5
Gut only	12	6
Neither	4	3
Total	27	17

Loss of Virus from the Mirid and Possible Transmission Routes

The results described so far do not explain how virus is transmitted to the plant by mirids. Inability to detect VTMoV in the salivary glands by ELISA has failed to implicate them in transmission and transmissibility appears to be independent of the presence or absence of virus in the hemolymph. To study the possible transmission routes from the mirid to the plant, experiments were done to determine the rate of loss of virus from the gut. In addition, experiments were done to determine whether virus in the gut was infective and whether there were mechanisms for inoculation of virus-contaminated mirid secretions to plants.

In an experiment to determine the rate of loss of virus from the mirid, nymphs were given acquisition periods of either 100 or 3000 min, and were then placed on the immune tomato host. Every 2 days for the next 9 days, 10 mirids were removed and their gut assayed for virus using the HRP assay. There was a linear relationship between the percentage of mirids with virus in the gut and time after acquisition. Although there was no effect of acquisition time on the rate of loss of virus from the gut, more nymphs contained virus in the gut if they had acquired virus for a longer period (Fig. 1.6). These results are consistent with the observed loss of virus from mirids over 9 days as determined by ELISA (9).

To determine whether fecal samples from viruliferous mirids was infective, mirids were given access to virus and then kept on tomato plants. Fecal samples were collected from infective nymphs 1 day after acquisition

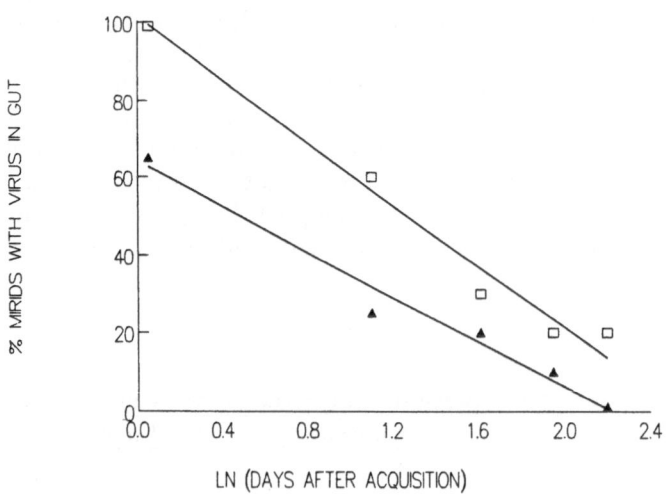

LN (DAYS AFTER ACQUISITION)

FIGURE 1.6. Relationship between days after acquisition (LN days) and percentage of mirids with virus in their gut. Two different acquisition access periods were tested, 100 min (▲) and 3000 min (□). Lines were fitted by least-squares regression with $r^2 = 0.98$ and 0.97 for top and bottom lines, respectively.

by gently squeezing their abdomens and collecting fecal drops on a square of parafilm. Fecal samples were infective when tested by mechanical inoculation onto healthy test plants. At 6 days after acquisition, fecal samples were collected from different mirids within the same experiment and these samples were also infective when inoculated onto test plants. Thus virus could remain infective in feces for at least 6 days after acquisition (10).

An experiment was done to determine whether purified virus or infective plant sap applied to the surface of a leaf could be inoculated by mirids and whether feeding was required. Infective sap or purified virus was used to simulate infective feces in these experiments (10). Noninfective nymphs given prior access to purified virus or infective sap painted onto a healthy leaf surface, transmitted VTMoV when transferred to healthy test plants. No transmission occurred if the nymphs first had their labia excised (10). Thus, feeding is required for transmission of VTMoV, and virus contaminating a leaf surface can then be introduced into a plant by feeding (10).

Infective virus could be retained in feces for up to 6 days following acquisition, and mirids could inoculate virus after being given access to infective sap painted onto a healthy leaf surface. However, it was not known whether mirids leaving the source of virus normally defecated infective virus for periods that could account for transmission up to 9 days following acquisition. To test this, nymphs were fed black nigrosin dye through a membrane and the number of nymphs depositing black or green feces was recorded up to 8 days after acquisition. The pattern of defecation of dyed food in nymphs is shown in Table 1.5. Some nymphs deposited only black (nigrosin-containing) feces for up to 5 days and the number of such nymphs decreased with time. A mixture of black and green feces continued to be deposited for 6 days, and by 7 days green feces had replaced black feces (10). Defecation seemed to be variable and some nymphs defecated all green feces one day and then a mixture of black and green the next. This suggests that "first in" is not necessarily "first out" and has implications for the ability of mirids to transmit VTMoV nonsuccessively over a relatively long period of 9 days, if feces are indeed the source of virus infectivity.

TABLE 1.5. Numbers of nymphs depositing black or green feces after being fed nigrosin dye through a membrane.

	Days after acquisition							
	1	2	3	4	5	6	7	8
Color of feces								
Black	11	5	0	1	1	0	0	0
Black and green	0	3	2	2	1	1	0	0
Green	0	4	10	8	10	9	11	11
No feces	2	1	1	2	1	3	2	2

Conclusion

Summary of Results

Our studies have shown that transmission of VTMoV by *C. nicotianae* has characteristics of both a noncirculative (nonpersistent and semipersistent), and circulative association. Although the short acquisition period is characteristic of a noncirculative, nonpersistent association, the inoculation period of between 1 and 2 h, and persistence for up to 9 days is characteristic of either a semipersistent, noncirculative or a circulative association. The ability of mirids to transmit through a molt is taken as evidence that virus circulates in the mirid and this is supported by the evidence that mirids can transmit following injection of virus into the body cavity of mirids (9). Virus was not detected in the salivary glands and, therefore, VTMoV is considered to be translocated by its vector rather than circulated as the latter term is used currently to describe associations involving the salivary glands (9). Failure to detect virus in the salivary glands may indicate either that mirids transmit VTMoV by a mechanism involving translocation of virus and inoculation to plants in a way that does not involve the salivary glands, or that levels of virus in the salivary glands were below the limit of detection of the immunoassays used (10).

Virus was detected in whole mirids by ELISA up to 9 days after acquisition, and in the gut as well for this period using the HRP immunoassay. Virus was detected also in the hemolymph using this method. These results suggest that sap accumulates in the gut during feeding and that VTMoV translocates from the gut to the hemolymph. Very low levels of virus were detected in the hemolymph and, therefore, it is likely that the virus detected in the whole mirid assays was actually from the infective sap in the gut. Because the presence of virus in the gut or hemolymph did not guarantee transmission, it is not surprising that more mirids contained virus as detected by ELISA than could transmit.

Nevertheless, a mechanism does exist that allows nonsuccessive transmission for up to 10 days after acquisition, and we have shown that virus can remain in the gut for this period, possibly to supply virus to a putative transmission pathway (10). We have shown also that virus retained in the gut is infective, and that during normal activity mirids can defecate material ingested up to 6 days previously. Furthermore, we have shown that mirids can acquire virus after feeding on a healthy leaf that had been painted with infectious plant sap previously.

A Hypothesis to Explain Transmission

As stated, VTMoV transmission by *C. nicotianae* has some characteristics of noncirculative transmission, and an ingestion–egestion mechanism (13) may explain the short acquisition period and increase in transmission rates with increasing acquisition and inoculation times (8). Even retention of infectivity up to 10 days may be due to regurgitation of infectious plant sap from the gut to the mouthparts but this mechanism has not been tested (10).

Our results indicate that the persistence of mirid infectivity could be explained by an ingestion–defecation model of transmission. In this model, virus ingested in plant sap during feeding accumulates in the gut where it does not propagate but is retained for up to 9 days. Virus in the gut is eliminated only slowly and serves as a reservoir for long-term transmission. During normal activity, mirids may defecate infective sap up to 6 days after ingestion and once on the leaf surface, virus may be inoculated either by mirids ingesting infectious feces and then regurgitating them during probing, or by contaminating the mouthparts while probing (10).

Mirid Transmission is Similar to Beetle and Mite Transmission

Some characteristics of transmission by mirids are similar to those of beetle-transmitted viruses. Although there are no reports of transstadial passage between larva and adult in beetles, transmission is nonsuccessive during serial transfer experiments and adults can transmit after injection of virus into the hemocele (14). One model to explain beetle transmission proposes that viruses transmitted by beetles enters the hemocele and passes into the gut lumen and then into the exterior by regurgitation or defecation. Fecal contamination of plant areas damaged by feeding could lead to inoculation (14).

Mite transmission of wheat streak mosaic virus resembles the transmission of VTMoV by the mirid. Although the virus is circulative in mites and is inoculated mainly to plants via salivary secretions during feeding, other modes of transmission are possible (21). Virus that accumulated in the mite gut may flow back towards the mouthparts and be introduced into plant cells during feeding. Alternatively, due to the large concentration of virus particles in the hindgut, some of the virus might be defecated and introduced into the leaf through feeding punctures, as described for the mirid, or through the abrasive action of the anal sucker and setae (22).

The Future

We think that the mirid–virus association provides a useful model for studying systems that do not fit any one transmission category. This is particularly so because stylet-feeding rather than chewing allows the use of membrane-feeding techniques to fine-tune acquisition experiments and overcomes the problem of chewed leaf fragments being retained in the gut and affecting assay results. Further, mirids molt at regular intervals unlike beetles and mites, and are easy to manipulate.

Although the ingestion–defecation model does not explain all our results, it provides direction for further experiments, in particular, the significance of ingestion–egestion behavior in transmission and an ultra-structural study of mirid salivary glands to detect very low levels of virus. It is probable that as different insect families are studied as virus vectors, it may prove increasingly difficult to categorize transmission using the traditional characteristics derived from aphid and leafhopper studies. We may need to move toward more general flexible models and accommodate more than one model to explain a single virus–vector association. Our study has shown that at least two mechanisms, ingestion–egestion and ingestion–defecation, may be required to describe transmission of VTMoV by the mirid *C. nicotianae*, a previously unknown virus vector.

References

1. Caciagli, P. Roggero, P., and Luisoni, E., 1985, Detection of maize rough dwarf virus by enzyme-linked immunosorbent assay in plant hosts and in the planthopper vector, *Ann. Appl. Biol.* **107**:463–471.
2. Fargette, D., Jenniskens, M-J., and Peters, D., 1982, Acquisition and transmission of pea enation mosaic virus by the individual pea aphid, *Phytopathology* **72**:1386–1390.
3. Forster, R.L.S. and Jones, A.T., 1980, Lucerne transient streak. *Commonwealth Mycological Institute/Association of Applied Biologists Descriptions of Plant Viruses* No. 224.
4. Francki, R.I.B., Randles, J.W., and Graddon, D.J., 1988, Subterranean clover mottle virus, *Association of Applied Biologists Descriptions of Plant Viruses* No. 329.
5. Fye, R.E., 1984, Damage to vegetable and forage seedlings by overwintering *Lygus hesperus* (Heteroptera: Miridae) adults. *J. Econ. Entomol.* **77**:1141–1143.
6. Gagne, S., Richard, C., and Gagnon, C., 1984, Silvertop diseases of grasses: state of knowledge, *Phytoprotection* **65**:45–52.
7. Getzin, L.W., 1983, Damage to inflorescence of cabbage seed plants by the pale legume bug (Heteroptera: Miridae). *J. Econ. Entomol.* **76**:1083–1085.
8. Gibb, K.S. and Randles, J.W., 1988, Studies on the transmission of velvet tobacco mottle virus by the mirid, *Cyrtopeltis nicotianae. Ann. Appl. Biol.* **112**:427–437.
9. Gibb, K.S. and Randles, J.W., 1989, Non-propagative translocation of velvet tobacco mottle virus in the mirid *Cyrtopeltis nicotianae. Ann. Appl. Biol.* **115**:11–15.
10. Gibb, K.S. and Randles, J.W., 1990, Distribution of velvet tobacco mottle virus in its mirid vector and its relationship to transmissibility. *Ann. Appl. Biol.* in press.
11. Gibbs, A.J. and Harrison, B., 1976, Plant Virology: The Principles, Edward Arnold, London, 292 p.
12. Greber, R.S. and Randles, J.W., 1986, Solanum nodiflorum mottle virus, *Association of Applied Biologists Descriptions of Plant Viruses* No. 318.
13. Harris, K.F., 1977, An ingestion-egestion hypothesis of noncirculative virus

transmission Harris, K.F., and Maramorosch, K. (ed): in *Aphids as Virus Vectors*, Academic Press, New York and London, pp. 165–220.

14. Harris, K.F., 1981, Arthropod and nematode vectors of plant viruses. *Annu. Rev. Phytopathol.* **19**:391–426.

15. Hawkes, R., Niday, E., and Gordon, J., 1982, A dot-immunobinding assay for monoclonal and other antibodies. *Anal. Biochem.* **119**:142–147.

16. Hori, K., Hashimoto, Y., and Kuramochi, K., 1985, Feeding behaviour of the timothy plant bug, *Stenotus binotatus* F. (Hemiptera: Miridae) and the effect of its feeding on orchard grass. *Appl. Entomol. Zool.* **20**:13–19.

17. McClintock, J.A. and Smith, L.B., 1918, True nature of spinach-blight and relation of insects to its transmission. *J. Agricul. Res.* **14**:1–59.

18. Matthews, R.E.F., 1981, Plant Virology, 2nd ed., Academic Press, New York and London, 897 p.

19. Morill, W.L., Ditterline, R.L., and Winstead, C., 1984, Effects of *Lygus borealis* Kelton (Hemiptera: Miridae) and *Adelphocoris lineolatus* (Goeze) (Hemiptera: Miridae) feeding on sainfoin seed production. *J. Econ. Entomol.* **77**:966–968.

20. Moshy, A.J., Leigh, T.F., and Foster, K.W., 1983, Screening selected cowpea *Vigna unguiculata* (L.) Walp., lines for resistance to *Lygus pratensis* (Heteroptera: Miridae). *J. Econ. Entomol.* **76**:1370–1373.

21. Paliwal, Y.C., 1980, Fate of plant viruses in mite vectors and nonvectors. Harris, K.F., Maramorosch, K. (ed): in Vectors of Plant Pathogens, Academic Press, New York and London, pp. 357–373.

22. Paliwal, Y.C., and Slykhuis, J.T., 1967, Localisation of wheat streak mosaic virus in the alimentary canal of its vector *Aceria tulipae* Keifer, *Virology* **32**:344–353.

23. Randles, J.W., and Francki, R.I.B., 1986, Velvet tobacco mottle virus, *Association of Applied Biologists Descriptions of Plant Viruses* No. 317.

24. Randles, J.W., Davies, C., Hatta, T., Gould, A.R., and Francki, R.I.B., 1981, Studies on encapsidated viroid-like RNA. 1. Characterisation of velvet tobacco mottle virus, *Virology* **108**:111–122.

25. Stahl, F.J. and Luepschen, N.S., 1977, Transmission of *Erwinia amylovora* to pear fruit by Lygus spp. *Plant Dis. Report.* **61**:936–939.

26. Tamada, T. and Harrison, B.D., 1981, Quantitative studies on the uptake and retention of potato leafroll virus by aphids in laboratory and field conditions. *Ann. Appl. Biol.* **98**:261–276.

27. Walstrom, R.J., 1983, Plant bug (Heteroptera: Miridae) damage to first-crop alfalfa in south Dakota. *J. Econ. Entomol.* **76**:1309–1311.

28. Woodward, T.E., Evans, J.W., and Eastop, V.F., 1979, Hemiptera, in *The Insects of Australia*, The Division of Entomology, Commonwealth Scientific and Industrial Research Organization, Melbourne University Press, Melbourne, pp. 387–457.

2
Factors Influencing Aphid Population Dynamics and Behavior and the Consequences for Virus Spread

Nick Carter and Richard Harrington

Introduction

Aphids are major pests of most temperate crops, causing direct damage through feeding and also, perhaps more importantly, as vectors of viruses. Virus diseases spread by aphids are classified into three main groups according to their relationship with their vectors: (1) persistent; (2) semipersistent; and (3) nonpersistent. A persistent virus can be acquired by an aphid feeding on an infected plant after about 20 min, although acquisition is more likely to occur after several hours. Acquisition is followed by a latent period before the virus can be inoculated to another plant. The aphid then usually remains infective for the remainder of its life. The virus either may replicate (propagative) or not within the aphid. Propagative viruses are sometimes transmitted directly to the offspring of the infected vector and all persistent viruses are retained through molts. A nonpersistent virus can be acquired by an aphid probing on an infected plant for only a few seconds, or minutes, whereas probes of longer duration often decrease the efficiency of transmission. Acquired viruses can be inoculated into a plant as soon as aphids probe but are retained only for a few hours. Nonpersistent viruses are carried on or inside the aphid mouthparts and never replicate within an aphid. The mechanism of transmission is still not fully understood, especially that of virus attachment to, and release from, the aphid. Some potyviruses need a "helper component" (41, 42) to be transmitted. This is a viral gene product that

N. Carter, Department of Entomology and Nematology, AFRC Institute of Arable Crops Research, Rothamsted Experimental Station, Harpenden, Hertfordshire, AL5 2JQ, England, U.K.
R. Harrington, Department of Entomology and Nematology, AFRC Institute of Arable Crops Research, Rothamsted Experimental Station, Harpenden, Hertfordshire, AL5 2JQ, England, U.K.

TABLE 2.1. Classification of viruses discussed in the text.

Transmission characteristics	Virus	Abbreviation
Persistent		
	Barley yellow dwarf	BYDV
	Beet mild yellowing	BMYV
	Beet western yellows	BWYV
	Lettuce necrotic yellows	LNYV
	Pea enation mosaic	PEMV
	Potato leaf roll	PLRV
	Sowthistle yellow vein	SYVV
Semipersistent		
	Beet yellows	BYV
	Strawberry yellows	SYV
Nonpersistent		
	Bean yellow mosaic	BYMV
	Cabbage mosaic	CMV
	Henbane mosaic	HMV
	Maize dwarf mosaic	MDMV
	Papaya mosaic	PaMV
	Pea mosaic	PMV
	Potato virus Y	PVY
	Soybean mosaic	SMV
	Sugar beet mosaic	SBMV
	Watermelon mosaic	WMV1

probably acts by binding the virus to a receptor site in the aphid. Not all potyviruses induce production of a helper component and some viruses can use those produced by other viruses. Semipersistent viruses have some transmission characteristics of nonpersistent and some of persistent viruses. Acquisition usually takes several hours but the virus is not circulated within the body of the aphid and there is no latent period. However, the aphid is able to transmit the virus over a period of several days but the virus is not retained through a molt. A few viruses do not fall into these three categories but have a bimodal pattern of acquisition (71). Aphids acquire such viruses during probes of several seconds or minutes and also during probes of several hours or days but with a decrease in efficiency with intermediate periods. Aphids transmitting after acquisition in the first peak have a nonpersistent relationship with the virus, whereas those in the second peak have a semipersistent one. No bimodally transmitted viruses are considered in this chapter.

Factors that influence the behavior and population dynamics of aphids will differ in their influence on the spread of the different types of virus. This review considers the effects of abiotic and biotic factors on aphid population dynamics and behavior and the consequences for the spread of viruses, particularly those viruses causing diseases in arable crops (Table 2.1).

The Viruses

Barley yellow dwarf virus (BYDV) is a group of persistent luteoviruses that do not replicate within an aphid and are not passed directly from mother to offspring although they are likely to acquire it from the plant the mother has infected. Barley yellow dwarf (BYD) is a widespread damaging disease caused by several isolates of at least two distinct viruses (85). These isolates are named after their principal vectors: MAV, *Sitobion (Macrosiphum) avenae* virus; RPV, *Rhopalosiphum padi* virus; RMV, *Rhopalosiphum maidis* virus; SGV, *Schizaphis graminum* virus; and PAV, *R. padi* and *S. avenae* virus (91). The life cycles of these aphid vectors range from asexual overwintering on cereals (most BYDV vector species seem to have the capability of doing this) to sexual overwintering as eggs (e.g., *R. padi* on bird cherry, *Prunus padus*) (19). The latter type of life cycle may be monoecious (non-host-alternating, e.g., *S. avenae*) or heteroecious (host-alternating, e.g., some *R. padi*). These differences in aphid life cycles, together with variations in the phenology and size of migrations are key components in determining the relative dispersal and incidence of the various isolates.

Potato leaf roll virus (PLRV) is also a luteovirus transmitted in the persistent manner and is probably nonpropagative in the aphid. It is confined mainly to the phloem tissue of host plants but may spread into epidermal and parenchymal cells in association with other viruses (7). Aphids must feed for at least 1 h in order to acquire the virus, and there is a latent period of approximately 12 h before an aphid can become infective. Aphids then remain infective for life. Most known hosts are Solanaceae although one potentially important host is *Capsella bursapastoris* (Cruciferae), which is also a host of the potato-feeding vectors *Myzus persicae* and *Macrosiphum euphorbiae*. There are several isolates of differing virulence and ease of transmissibility. In Western Europe, the virus is confined largely to potato and is perpetuated via the seed tubers. Its economic importance is largely in the seed-growing industry as primary symptoms are not severe generally.

Potato virus Y (PVY) is a potyvirus transmitted in the nonpersistent manner by many colonizing and transient species (11). Optimal acquisition feeding time is typically 15 to 60 s, with those periods lasting more than 2 min being less favorable. Most aphids that are feeding lose the ability to transmit within an hour of acquisition but starved aphids can retain the virus for longer. It can be transmitted also mechanically. It resides principally in the epidermal cells and occurs mainly in the Solanaceae. There are several groups, each consisting of different strains. The PVY[O] group occurs worldwide and induces leaf crinkling or mottling in potato grown from infected seed tubers or true seed, although the severity of the symptoms differs between cultivars. Strains of PVY[N] occur in Europe, parts of Africa, and in South America and will induce mild mottling in potato grown from infected seed. Other less economically important groups also exist.

Beet yellows virus (BYV) is a closterovirus transmitted by many aphid species, but especially *M. persicae*, in the semipersistent manner. It can be sap-inoculated but with difficulty. It is widespread throughout the beet-growing areas of the world and is confined mainly to the Chenopodiaceae. There are several strains that differ in the severity of symptoms they cause. The virus typically has a half-life of approximately 8 h in the vector and is retained for up to 3 days. It requires acquisition feeds of more than 12 h and inoculation feeds of at least 6 h for maximum transmission. There is no latent period in the aphid and vectors do not retain the virus after molting.

Beet mild yellowing (BMYV) another virus causing yellowing of beet, is a luteovirus with transmission characteristics similar to those described for PLRV.

Weather

Overwintering Survival

In temperate regions, the weather during the winter is an important factor influencing aphid survival and hence the numbers and time of migration of the vectors in the spring (115).

The effects of winter weather vary depending on the proportion of holocyclic to anholocyclic clones in the population. Holocyclic clones are less affected by severe weather conditions as they overwinter as eggs, which are relatively cold-hardy. Consequently, species that are largely anholocyclic usually show a more significant correlation between winter temperatures and the number of alatae flying in spring (Table 2.2). Thus the effects of winter weather on virus spread are influenced by the biology of the vectors rather than by the transmission properties of the virus.

M. persicae, an important vector of PVY, PLRV, BYV, and BMYV, is mainly anholocyclic in southern England, and the time of the first appearance of alatae in suction traps and the number sampled in the spring are correlated significantly with temperature in January and February (Fig. 2.1) (6). Temperatures during the winter are correlated with the incidence of sugar beet yellows (caused by BYV and BMYV) the following summer (Fig. 2.2) (55, 130). However, in the west of Britain the relationship between the time of the first suction trap record of *M. persicae* and winter temperature is weaker (6) and, consequently, the relationship between the incidence of sugar beet yellows and the mean temperatures during the winter is also weaker (130). In the west, rainfall and leaf wetness are higher and limit winter survival of *M. persicae*, rather than do low temperatures

TABLE 2.2. Correlation between mean screen temperature[a] and number of aphids sampled in the Rothamsted 12.2-m suction trap[b]

Aphid Species and Life Cycle	r	p
Largely anholocyclic species		
Myzus ascalonicus	0.66	< 0.001
Macrosiphum euphorbiae	0.67	< 0.001
Myzus persicae	0.69	< 0.001
Sitobion avenae	0.54	< 0.01
Mixed anholocyclic/holocyclic species		
Acyrthosiphon pisum	0.19	NS
Brevicoryne brassicae	0.59	< 0.01
Cavariella aegopodii	0.00	NS
Rhopalosiphum padi	0.50	< 0.05
Largely holocyclic species		
Brachycaudus helichrysi	0.56	< 0.01
Metopolophium dirhodum	0.20	NS
Aphis fabae	0.15	NS
Phorodon humuli	0.11	NS
Hyperomyzus lactucae	0.51	< 0.05
Drepanosiphum platanoidis	−0.36	NS

[a] December to February inclusive.
[b] Up to June 30 inclusive.

(54). Overwintering of this species is most successful in the southeast where conditions are relatively mild and dry (114).

The significant relationship between the time of the first alate *M. persicae* in the suction trap at Edinburgh with the mean screen air temperatures in February has been used to advise growers of the need to apply insecticide granules at planting to control vectors of PLRV in the valuable Scottish seed potato crop (118, 119). In years when the forecast indicates the start of migration before the middle of June, insecticide use at planting is justified but if it is predicted to occur later, treatment at planting is not recommended as the granules are not persistent enough to control the aphids arriving later and, thus, sprays may be necessary.

Laboratory studies have shown that a temperature of −11°C over 72 h kills 50% of adult *M. persicae* acclimated (reared in the laboratory at low temperatures for one generation) at 5°C but even lower temperatures are needed to kill first instar aphids or older aphids exposed for a shorter period (6). Although higher temperatures kill fewer aphids directly, they may inhibit movement. This may still result in death of aphids if they are feeding on senescing leaves, as these will be shed during the winter (54). Growth and reproduction of *M. persicae* can continue in the field even when the mean temperature is as low as 0°C (54). In a study of movement in *S. avenae*, all instars were capable of movement at 3°C, whereas at −1°C no first instar aphids could move. None of the instars could move at −4°C (101). Thus below these temperatures virus spread is minimal.

Spring-sown cereal crops that are sown later than average crops are at

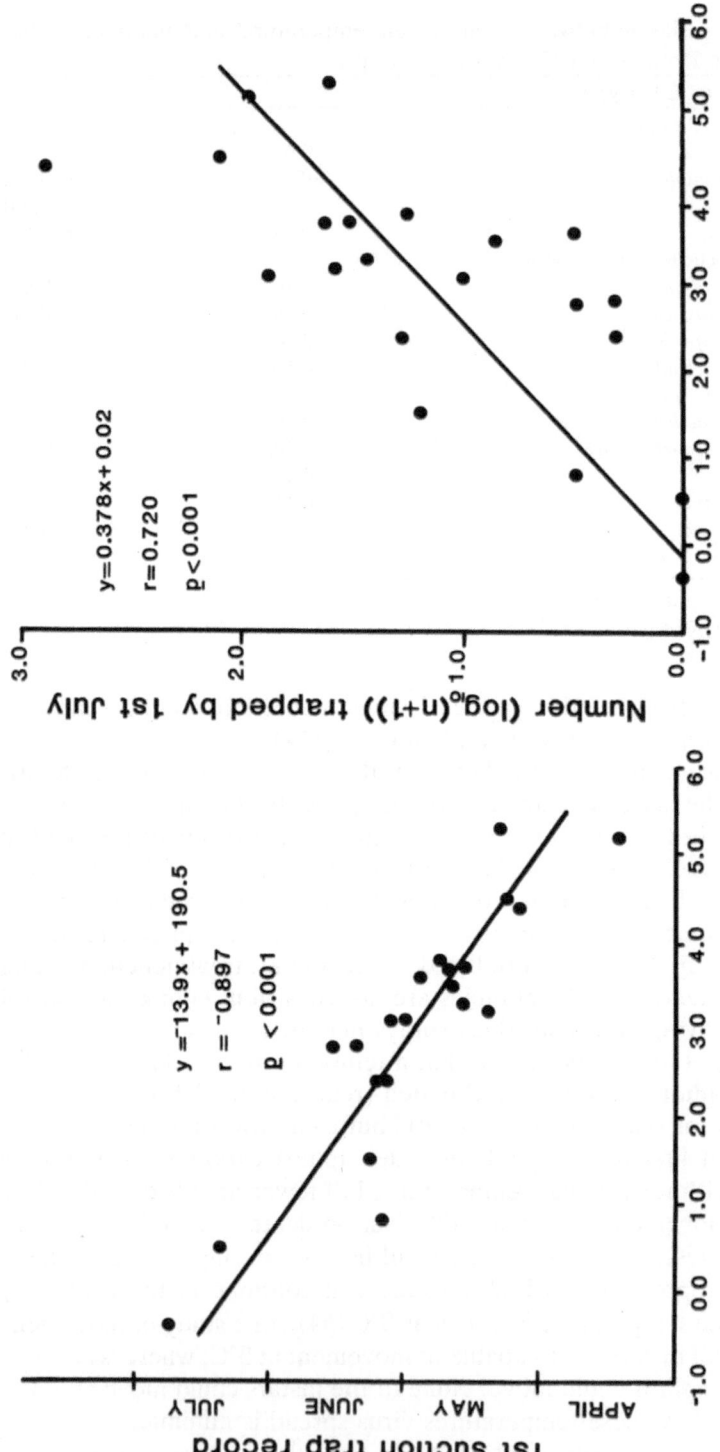

FIGURE 2.1 The effect of temperature during the winter on the date of the first *Myzus persicae* in the Rothamsted 12.2-m suction trap and the numbers sampled in the spring and early summer, 1966–1988.

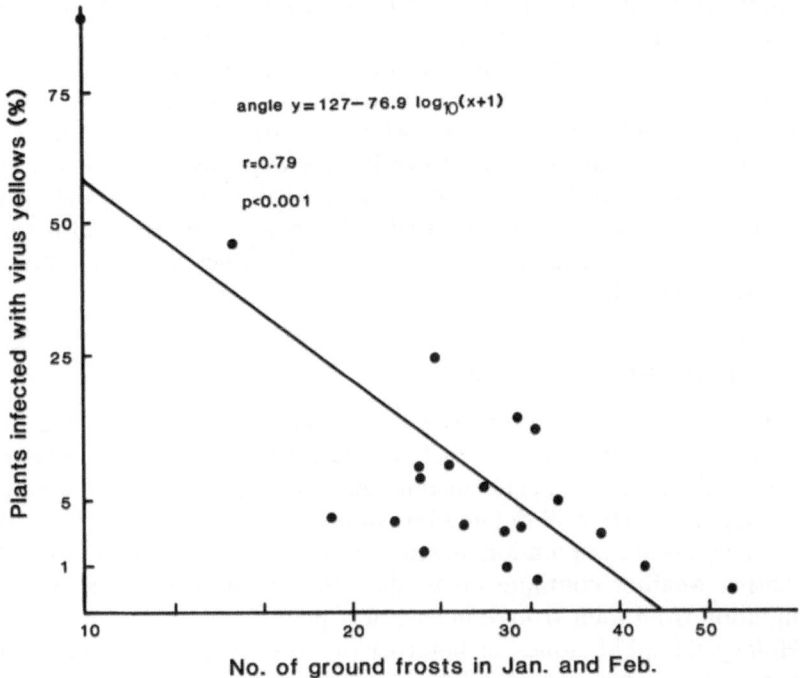

FIGURE 2.2. The effect of temperature during the winter on the incidence of beet yellowing viruses (angle scale) at the end of August in Eastern England, 1965–1985.

more risk from infection with BYDV than those sown early, as they are exposed to the spring aphid migration just after emergence when they are most susceptible to damage from the disease. However, large numbers of alatae of *S. avenae* occur early in spring following mild winters (25, 125), which increases the risk of infection and yields loss even of early-sown cereals that normally escape significant damage. This occurred in the UK and in other west European countries in 1989 after the very mild winter resulted in the growth of cereal aphid populations on autumn-sown cereals during the winter and spring. These aphids then migrated to spring-sown cereals.

Winter weather affects aphid survival not only directly but also indirectly through its influence on host plants. Weeds provide an important overwintering site for many anholocyclic aphid clones and aphid survival depends as much on the cold-hardiness of their host plants as on their own cold-hardiness (120). *M. persicae* feeds preferentially on small nettle, *Urtica urens*, which is not cold-hardy, whereas *M. euphorbiae* is more catholic and, hence, is usually more common in Scotland, particularly after a severe winter. Both species are important vectors of potato viruses, and although

M. persicae is the more efficient vector of PLRV and PVY, *M. euphorbiae* may be more important because it survives in larger numbers.

Root and sugar beet clamps can be important sources of *M. persicae*, other vectors of sugar beet viruses, as well as of the viruses themselves (14, 56). Clamps can be much warmer below the surface than is the air outside (14) and, hence, can provide protection for overwintering aphids. If not cleared before sugar beet is planted, clamps can increase the risk of virus infection in adjacent beet crops greatly. Ornamental flowers, lettuces, and vegetables grown in glasshouses and heated frames are other potential local sources of virus vectors (34).

Spring Population Development

For holocyclic clones, spring can be considered to begin at egg hatch, except for anomalous species such as *Brachycaudus helichrysi*, where the eggs hatch in autumn. The distinction between winter and spring for anholocyclic clones is ill defined but spring will be considered here as the start of significant population increase. For both types of clone, the effects of spring weather continue up to the time aphids land on a crop after emigration from their overwintering host plants.

Holocyclic aphid clones of heteroecious species produce one or more generations on the winter host plant before production of alatae that migrate to the summer host. Clones passing the winter as active stages do not have this time delay prior to emigration, and alatae may fly as soon as it becomes warm enough to do so. Consequently, in warm springs, crops are likely to be infested earlier, which increases the risk of virus spread. Population development is influenced mainly by temperature but the migration to crops is affected by windspeed and rainfall also.

Temperature affects the rate of build-up of aphid populations in spring and subsequent virus incidence for a number of persistent, semipersistent, and nonpersistent viruses.

Higher temperatures through winter and spring, over 4 years, increased spread of PLRV and PVY in Poland (35) but the broad time periods examined did not allow separate consideration of effects on winter mortality and spring population development. However, as all the vectors overwinter as eggs in Poland, it seems likely that the effect is predominately on spring population increase.

The incidence of sugar beet yellowing viruses in England in August is related not only to temperatures in winter but also to those in April (130). Temperature data from Rothamsted correlated with virus incidence in all parts of the country more closely than with data from separate regions where virus incidence was recorded. It was suggested that Rothamsted lies in the path of winds from various areas where *M. persicae* most successfully overwinters and is, therefore, in a strategic position to monitor temperature conditions relevant to virus spread. This seems unlikely, however, as

M. persicae has been recorded rarely in suction traps in April in any part of England. April temperatures are more likely to affect population increase and, hence, the number of migrants. As Rothamsted is on the edge of the area of most successful overwintering (114), April temperatures there may affect available migrants more than do temperatures in the main beet growing areas. However, the relative contribution of local populations and those from further afield in virus spread is poorly understood. It may be misleading to differentiate migrants into local and long distance. It is more likely that the distances that aphids fly are on a continuum with no sharp distinction between the two types of migrant (115).

As it is very difficult to obtain long runs of weather data that represent large areas of arable land, the usefulness of Rothamsted data may be a reflection of the unsuitability of other weather stations used in the initial analysis. Indeed, reanalysis of the data for the forecast of beet yellowing viruses showed that temperatures at Broom's Barn, at the center of the main beet growing area of Eastern England, gave the most significant correlations with observed virus incidence in that region (55). In this new analysis, temperatures in January and February are more significant than those in the spring. Turl (119) also found that winter temperatures in Scotland are more important as a predictor of early aphid populations than are temperatures in spring, although temperatures during April are important in determining the extent of spread of PLRV by *M. persicae* and *M. euphorbiae* (60). *M. persicae* flew earlier and in greater numbers in Ontario (where it is holocyclic) in 1976, when the mean April temperature was much higher than usual, compared with temperatures of 1971 to 1975 when April was cooler (32). Also the previous November was much warmer than usual, and overwintering populations were favored by warm weather during the time *M. persicae* mates and oviposits. No virus incidence data are given but high incidence of nonpersistent viruses transmitted by *M. persicae* to capsicum peppers in the same area is correlated with high April temperatures and sunshine hours (Fig. 2.3) (65). In France, there is a positive correlation between temperature in the latter two-thirds of April and numbers of *M. persicae* and *Aulacorthum solani* sampled over potatoes during emigration from winter hosts in May (89).

Dispersal

The initiation of flight by aphids is dependent on many factors and there is much variation in behavioral responses between individual aphids to these factors (63, 116). There are however some conditions under which no aphids can fly. If these conditions persist for long, emigration is delayed (126), and many migrants may die on the overwintering host. As holocyclic clones produce emigrants more synchronously than do anholocyclic clones, adverse conditions for flight could be more catastrophic for the former if these conditions occurred at a critical time. However, as holocyclic clones

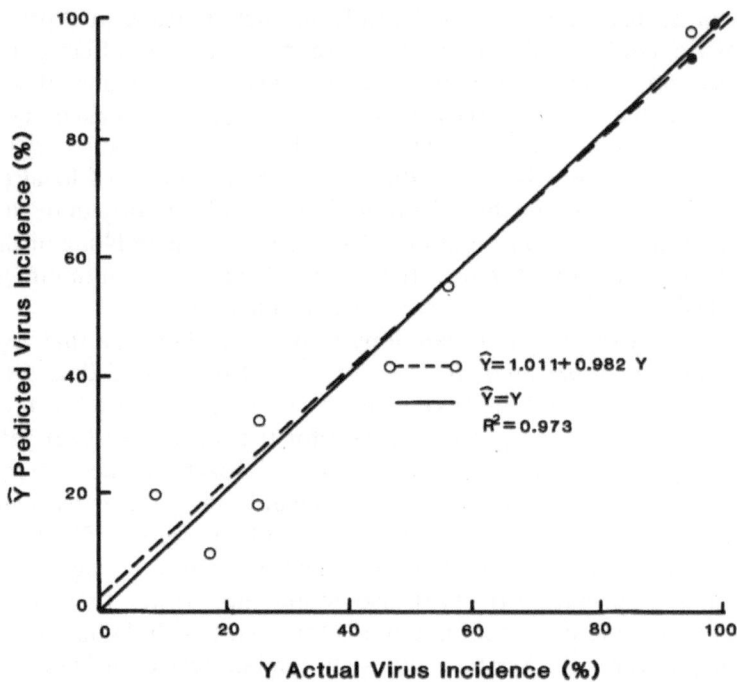

FIGURE 2.3. The relationship between predicted (Ŷ) and actual (Y) incidence of nonpersistent viruses in peppers (%) in Canada, based on daily day-degrees plus sunshine hours in April, 1970–1977. (From Ref. 65.)

of a particular species produce alatae later than do anholocyclic clones of the same species, adverse weather conditions are less likely to occur during the critical flight period. No work has been done that relates virus incidence to the timing of spring flight in isolation from the effects of the various stimuli on population development.

An individual aphid will not fly unless conditions are between lower and upper thresholds for temperature, light intensity, windspeed, relative humidity, and rainfall (28, 48, 137). These thresholds vary between individuals and between the same individual on different occasions. The climatic factors influencing take-off must be measured in the crop. The crop canopy may provide enough protection to permit take-off when the windspeed recorded by standard meteorological stations would suggest that it is unlikely. An average windspeed of 20 kmph at 10 m is equivalent to an average of 6.5 kmph in a potato crop canopy (12), which is similar to the take-off threshold for *M. persicae* (22). High windspeeds outside the crop canopy do not cause alatae, once airborne, to cease flying.

The minimum temperature for flight of *M. persicae* can be as low as 13°C (13), and above that there is a sigmoidal increase in the proportion flying until at 18°C 100% are airborne (56, 58). Taylor (113) found that no *Aphis fabae* were caught in suction traps unless the maximum temperature reached 15°C. Above that temperature, there was a sigmoidal increase in the proportion flying such that on occasions when the temperature reached 17.5°C, aphids were present in 50% of the samples, taken every 30 min, whereas at 21°C they were present in all of the samples.

The direction, duration, and strength of wind at the sources of virus vectors influence virus incidence away from these locations. Long-distance migration is probably more significant for persistent rather than for nonpersistent viruses as the latter are not retained sufficiently long in aphids. For example, the more extensive distribution of BMYV compared with BYV may be due partly to its greater persistence (94). Wellington (135) summarized the weather patterns likely to result in the establishment of new colonies of insects in sufficient numbers to be of economic importance, at both short and long distances from their source populations. It was concluded that for long-distance migrations to occur, wind patterns would need to combine constancy of direction with a lengthy duration. Such conditions are relatively uncommon so that long-distance distribution by wind is usually unimportant. Only occasionally would sufficient individuals of a given species be deposited in a suitable habitat. Short-range migrations that are influenced by such local topographical features as valleys are likely to result in the frequent deposition of individuals in the same areas and, hence, be more significant economically than would long-distance migrations.

Some outbreaks of nonpersistent viruses far from known sources, however, have been related to long-distance transport by winds. Zeyen, Stromberg & Kuehnast (141) produced circumstantial evidence that Maize dwarf mosaic (MDMV) is transported to Minnesota from the southern plains of the United States. Their studies showed that the virus could be retained by aphid vectors for 19 h, a period uncharacteristically long for nonpersistent viruses. In 1977, severe outbreaks of the disease occurred over large areas of Minnesota where infection is unusual. MDMV is endemic in the states of the southern plains,and in spring 1977 temperatures were unusually high resulting in larger populations than usual. However, a severe drought followed, causing a deterioration in crop condition that favored production of alatae. Toward the end of June southerly low-level jet winds formed between the southern plains and Minnesota, where a cold front caused thunder storms and prevented further migration. This resulted in a correlation between the amount of rainfall and the distribution of fields infected with MDMV.

In southwest Florida the distribution of Watermelon mosaic virus (WMV1) (also known as Papaya Ringspot Virus), which is nonpersistent, is related to the distribution of infected weeds around the edges of fields (2). Not surprisingly, infection occurs earlier and is more extensive in crops downwind from virus sources as aphids can fly against winds of less than approximately 2.4 kmph only (46).

Low-level jet winds have been associated with the spread of BYDV by *S. graminum* in Wisconsin (77). In 1959, late planting of oats, due to an unusually cold and wet April, meant that crops were young and favorable for aphid multiplication in May and June at the time when airborne populations were arriving. Hodson & Cook (59) reported *S. graminum* in adjacent Minnesota at the same time and attributed their arrival to winds carrying them from states further south where large populations had developed. In Iowa, BYDV incidence is related also to population levels of viruliferous vector aphids in the southern plains, the incidence of low-level jet winds, and the weather conditions in cereals after colonization (high temperatures encouraging the spread of disease and high rainfall limiting it) (123, 124).

In Wales, the numbers of *R. padi* and *R. insertum* sampled in a suction trap in September and October are correlated with preceding rainfall during the summer (both species) and accumulated day-degrees above 6°C in September and October (*R. padi*) or below 6°C in February (*R. insertum*) (1). Rainfall was thought to influence aphid population development indirectly by its effect on grass growth but the temperature during September and October is likely to affect the ability of aphids to migrate directly. Warm autumns are likely to result in large numbers of *R. padi* migrating, especially after a wet summer.

In Sweden, the incidence of sugar beet yellowing viruses is predicted from the proportion of winds in July that come from the southeast, south, or southwest, and from their speed (Fig. 2.4) (136). *M. persicae* probably

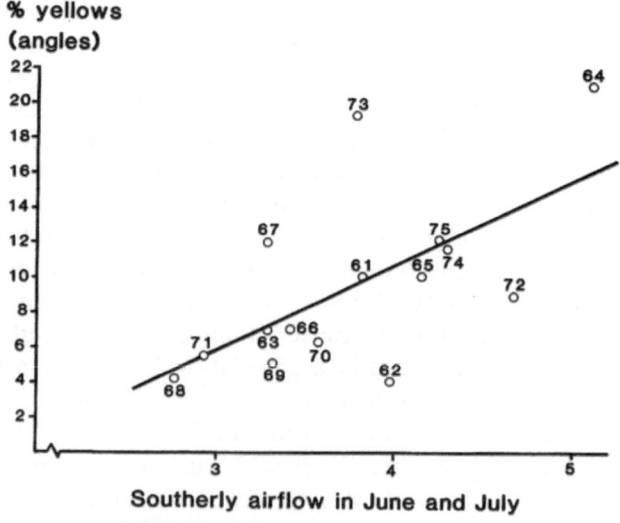

FIGURE 2.4. The relationship between percentage beet yellowing viruses (angles) in Sweden, and southerly airflow in June and July. (From Ref. 136.)

does not overwinter anholocyclically in Sweden and southerly winds were thought to bring viruliferous alatae from the European continent.

Secondary Spread

Most of the studies on the effects of weather on aphid population development on summer hosts have not considered virus spread.

Aphid build-up is only important to virus spread if it results in the movement of aphids within the crop. Aphid species that form large, dense colonies of apterae are less important as virus vectors in comparison with those that produce equivalent numbers that disperse. This is probably one reason why *M. persicae* is more important in the secondary spread of sugar beet viruses than is *A. fabae* (64). Although large colonies of sedentary species are likely to produce alatae eventually, this often happens when the plants are mature and less susceptible to damage from viruses. Also, the alatae are likely to leave the crop in search of less crowded or more nutritionally favorable host plants.

Alatae of transient and colonizing species are probably more important than resident apterae in the secondary spread of nonpersistent viruses. A colonizing alata is likely to probe several plants before finally settling to feed and reproduce. There are potentially large numbers of transient alatae that are able to infect susceptible plants during the summer as they move from plant to plant. Alatae of colonizing species, however, are more likely to cause secondary spread than are transient alatae as the latter probably emigrate from the crop after identifying it as an unsuitable feeding site.

Secondary spread of BYDV results in saucer-shaped discolored areas in crops caused by the local movement of aphids from initial central foci (84). Much of this movement must occur during the early winter in autumn-sown crops but the weather factors that influence it have not been quantified.

The effect of temperature on the spread of lettuce necrotic yellows (LNYV) by *Hyperomyzus lactucae* is better understood. The peak incidence of LNYV occurs 4 to 5 weeks after the peak numbers of *H. lactucae* because of the time taken for an infected aphid to become infective and for lettuce to show symptoms. Alate *H. lactucae* were trapped only when the mean weekly temperature was between 15 and 22°C, and most aphids were caught when the weekly windrun was between 1280 and 1920 km and rainfall was low (86). *H. lactucae* does not colonize lettuce but acquires the virus from sowthistle, *Sonchus oleraceus*, its main secondary host. An infected aphid only needs to probe on lettuce for 1 to 5 min in order to inoculate the virus, although it takes up to 30 min on sowthistle (10). In September and October temperatures are close to the optimum for aphid population development and this is when most virus transmission occurs.

To acquire or inoculate a persistent virus, an aphid must feed, whereas

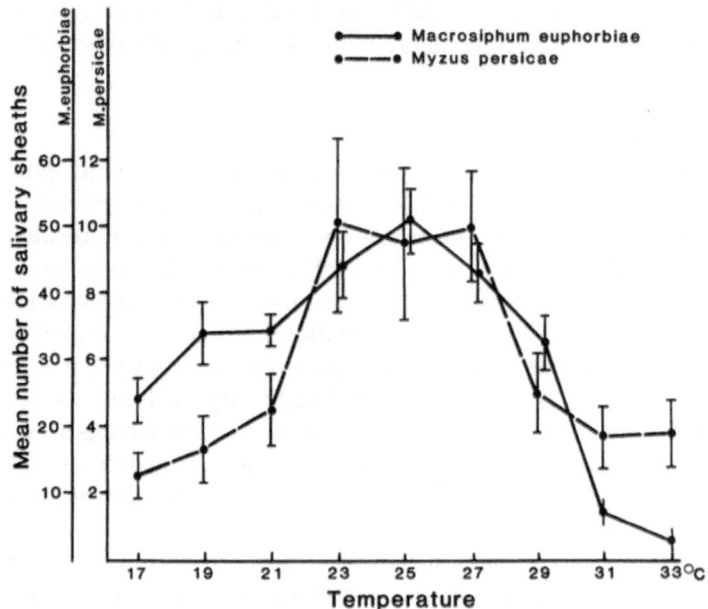

FIGURE 2.5. The effect of temperature on the number of salivary sheaths made by *Myzus persicae* and *Macrosiphum euphorbiae* in gels (±95% Confidence Intervals). (From Ref. 134.)

with a nonpersistent virus, the aphid only has to probe the host plant. Thus factors that affect feeding and probing will influence the efficiency of transmission of persistent and nonpersistent viruses, respectively.

Temperature affects the frequency (Fig. 2.5) (104, 134), the duration (97, 106, 127–129, 131), and the time taken for aphids to begin probing (106). Probing frequency in clones of *M. persicae* and *M. euphorbiae* is maximal between 23 and 27°C, and the highest efficiency of transmission of PVY occurs at 25°C, suggesting a link between the two observations (134). A larger number of probes would be expected to increase the probability of acquiring a nonpersistent virus and there is some evidence that acquisition of persistent viruses is increased also. More potato plants were infected with PLRV if acquisition, inoculation, or both occurred at 26 rather than at 12°C (104). Presumably, increased frequency of phloem penetration and stylet retraction contributes to increased transmission. The duration of probing is longer at lower temperatures, and which also results in a decrease in transmission of nonpersistent viruses (97, 131). The proportion of probes by *M. persicae* lasting more than a minute increased above and below 21°C (106). This relates to the reduced efficiency of transmission of the nonpersistent cabbage mosaic virus (CMV) (also known as Turnip Mosaic Virus) at these temperatures, with maximum acquisition occurring between 15.6 and 26.7°C. Inoculation efficiency, however, continued to increase with temperature from 4.4 to 32.2°C. Also, the efficiency of inoculation increased by higher temperatures, which raised the frequency of probes (105).

Viruses are transmitted optimally at different temperatures with different virus–vector combinations and these differences are not related to a particular aphid clone or virus (97). Transmission of one strain of PVY by *M. persicae* was dependent on temperature, but transmission of another strain was not. *M. euphorbiae*, however, transmitted both strains independently of temperature. If feeding behavior is the sole cause for differences in transmission efficiency of the two strains, then transmission of both strains by *M. persicae* would be expected to depend on temperature. As this was not the case, other factors must control vector efficiency also.

Temperature influences the spread of persistent viruses by affecting the duration of the period between acquisition of a virus by an aphid vector and the time when that vector becomes infective (latent period). An increase in temperature reduces the latent period of pea enation mosaic virus (PEMV) in *Acyrthosiphon pisum* (Fig. 2.6) (108), sowthistle yellow vein virus (SYVV) in *H. lactucae* (29), LNYV in *H. lactucae* (10), BYDV in *S. avenae and R. padi* (121), and PLRV in *M. persicae* (110, 111).

Temperature also affects the time that a virus is retained by an aphid. Aphids carrying persistent viruses generally remain infective for life, so the retention period may be expected to be affected only by the longevity of the vector. This may be shorter at higher temperatures (23). However, *A.*

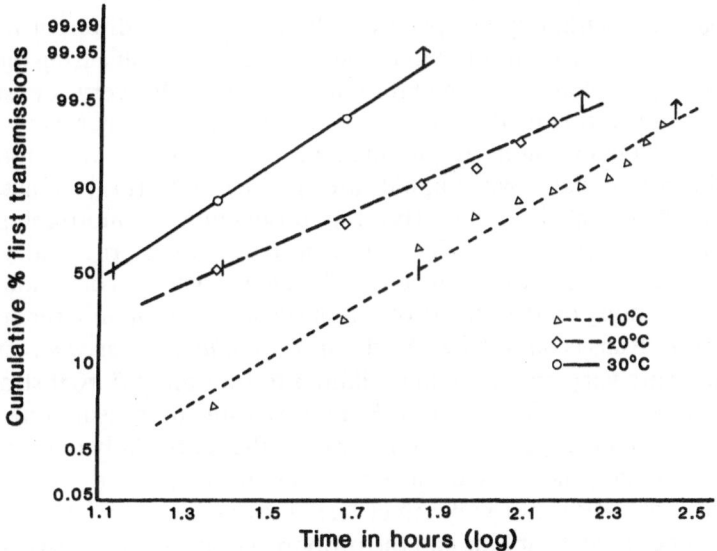

FIGURE 2.6. The relationship between latent period of pea enation mosaic virus and temperature for *Acyrthosiphon pisum* nymphs. (From Ref. 108.)

pisum does not always retain PEMV for life (108). The mean retention time declines with increase in temperature from 29.5 days at 10°C to 4.3 days at 30°C. The retention periods are slightly longer in alatae.

Semipersistent viruses are retained usually for longer than nonpersistent ones, but neither survive a vector molt. The semipersistent SYV is retained in *Chaetosiphon fragaefolii* for 3 to 4 days longer at 4.5°C that at 21°C, but no details are given of the actual periods nor how they relate to time of molts (78).

The retention times of nonpersistent viruses are short that the effects of temperature on their duration greatly affects the distance that the virus can be transported viably during aphid flight. Sugar beet mosaic (SBMV) henbane mosaic (HMV), and pea mosaic viruses (PMV) persist for longer at lower temperatures, and the percentage transmission drops at an average rate of 3% per degree-rise in temperature (21) so that few aphids transmit after 30 min at 30°C. The rate of decrease and its response to temperature are similar for flying or starved alatae and for starved apterae.

The loss of infectivity of CMV by *M. persicae* is more rapid at 30°C than at either 20 or 10°C, but similar at the two lower temperatures (106). Papaya mosaic virus (PaMV) is retained by *Aphis gossypii* for much longer at 4°C than at 20°C and for longer at 20°C than at 25 to 31°C (98).

Biotic Factors

Intrinsic Aphid Effects

Variation exists within aphid species in the responses to different factors. Therefore, generalizations of aphid biology based on results using one or a few aphid clones can be misleading but, unfortunately, very few studies have been done taking this variation into account. Clones vary in (i) adjustments to temperature requirements to suit local conditions (38), (ii) the production of alatae when aphids are crowded (69), (iii) production of sexual morphs (76, 109) and (iv) the production of morphologically recognizable biotypes (61). The latter has sometimes resulted in the breakdown of host-plant resistance as different biotypes become dominant (133). In one study on the effects of intraspecific variation on virus spread, clones from Kansas and New York of *R. maidis*, *S. avenae*, and *S. graminum* were very similar in their abilities to transmit different strains of BYDV, but a clone of *R. padi* from Kansas was able to transmit one strain that a clone from New York was not (92). As there are likely to be several clones in a region, these effects are not likely to influence the large-scale distribution of viruses but may affect local incidence.

The number of alate aphids and the proportion carrying the virus are two main factors influencing the amount of primary inoculation of persistent viruses introduced into crops.

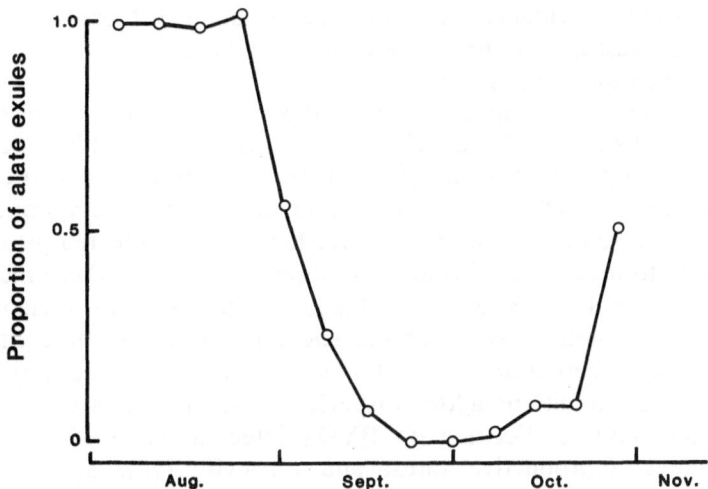

FIGURE 2.7. The proportion of alate female *Rhopalosiphum padi* that were alate exules sampled at 12.2 m in choice tests in 1986.

In Great Britain, cereal crops sown in September and early October are commonly infested with aphids carrying BYDV. *R. padi* is one of the most numerous aphid species caught in suction traps at this time (117), although these high aerial densities are not reflected in the size of the populations on crops (112). This is primarily because many of the female aphids sampled are gynoparae returning to the overwintering host (bird cherry, *P. padus*). Any BYDV that gynoparae introduce into crops is unlikely to be spread as they are reluctant to reproduce on cereals (27). Thus only the number of alate exules in the population is important in determining the epidemiology of BYDV. In 1986, using host plant choice tests, it was found that 5% of all *R. padi* trapped during the autumn, at 12.2 m at Rothamsted, were alate exules (Fig. 2.7) (112). This proportion is, however, likely to vary both annually and geographically. *S. avenae* is much less common in suction trap samples than *R. padi* during the autumn (usually 1:100 or less), but as it is monoecious on Gramineae, all alate morphs can colonize cereals. Thus the densities of the two species on crops are usually more similar than their aerial densities would suggest (25). As *R. padi* is less cold-hardy than is *S. avenae* (49) the latter species is usually more numerous in the following spring (25).

The proportion of alate aphids carrying BYDV, as measured in host-plant infectivity trials, is very variable. At two sites over 2 years in Victoria, Australia, the proportion of infected *R. padi* ranged from 23 to 61% (100). This is much greater than the range found in eastern England (1969–1973) of 1.1 to 11.5% (81), and of that in central Washington state, USA where the proportion ranged from 3.4 to 14.5% estimated from alatae collected in the field (47). It is not clear whether this is a result of differences in the incidences of BYDV in alternative host plants or because all clones in Australia are likely to be anholocyclic and thus there is no egg stage in which the virus is lost.

The production of alatae in cereal aphid colonies is influenced by density and host-plant stage, with more *S. avenae* alatae produced at high densities and at later crop growth stages (5, 132). In addition to these two factors, the presence of BYDV in host plants influences the production of alatae. Both *S. avenae* and *R. padi*, which have fed on virus-infected plants, are more likely to produce alatiform offspring compared with those that have fed on uninfected plants (39, 40). This is probably a result of changes in host-plant physiology (possibly amino acid concentrations), brought about by the virus, affecting aphid development. The alatae produced, if the environmental conditions allow, disperse the virus over larger distances than would apterae. Feeding on BYDV-infected plants also increases development, reproductive rates, and longevity (79), and decreases honeydew production (3). Similar observations have been made on *H. lactucae* feeding on SYV infected lettuces (107). These observations indicate a host-plant physiological response to the presence of the virus benefitting aphid feeding and population development.

Natural Enemies

Natural enemies attacking aphid populations can be classified into four groups: (1) hymenopterous parasitoids; (2) aphid-specific predators; (3) polyphagous predators; and (4) entomophagous fungi. Many techniques have been developed to measure their effect on aphid populations including correlations between densities of aphid and natural enemies, manipulative studies to increase or decrease natural enemy or aphid densities, and simulation modelling (9, 18, 139). The effectiveness of natural enemies in controlling virus spread is considered usually to be insignificant because damage can occur at low-aphid densities (18). In addition, because of the rapidity with which aphids can acquire and transmit nonpersistent viruses, natural enemies are likely to be of little value in their control. Thus relatively few studies have been done on the effects of changes in natural enemy density on virus spread as a consequence of differences in aphid population development.

Generally, it is accepted now that natural enemies can prevent cereal aphid populations developing to outbreak levels during the summer when they cause damage through feeding rather than as virus vectors. It is the combined effect of all four of the groups that is important (18). However recent enzyme-linked immunosorbent assays (ELISA) have shown that carabid and staphylinid beetles and spiders consume cereal aphids in the autumn, although subsequent virus incidence was not measured (102). Aphid populations were higher after application of the molluscicide methiocarb that resulted in greater spread of BYDV (68). No effect on polyphagous predator numbers extracted from soil samples was detected, although the increase in numbers of aphids and virus spread was attributed to a depletion in predator densities.

Jepson & Green (62) argued that as polyphagous predators are usually present prior to aphid immigration into sugar beet crops they are likely to be effective in reducing virus spread. In an earlier study, manipulation of polyphagous predator populations using ingress-only barriers and pesticide treatments in plots with complete barriers resulted in more aphids (mostly *A. fabae*) in plots with depleted predator populations, although the effects on virus incidence were insignificant (30). In contrast, Wratten & Pearson (140) demonstrated in New Zealand that depletion of the polyphagous predator population resulted in higher populations of *M. persicae* on sugar beet plants, more plants infested with aphids, and also higher virus incidence [beet western yellows virus (BWYV)]. There were, however, no effects on sugar yields as virus levels were thought to have been too low to cause damage.

Natural enemies may result in higher virus incidence if they disturb the aphids sufficiently for them to walk between plants. In a field-cage experiment, pea aphid, *A. pisum*, moved more between plants in the presence of adult coccinellids, *Coccinella californica*, which resulted in a higher incidence of BYMV (93). However, the relative importance of this stimulus compared with such other factors as the intrinsic urge for an aphid to move is unknown for most aphid virus vectors.

Alternative Host Plants

Viruses that infect annual crops often need alternative host plants in which to survive when crops are not present. Exceptions to this include seed- or tuber-borne viruses and those that infect such crops as rice, which have several overlapping growing seasons in a year. However, even when continuous propagation within a crop is possible, alternative host plants may still play an important role in epidemiology.

For BYDV, the major host plants between harvest of one cereal crop and emergence of the next are wild and cultivated grasses (although maize may be important in some areas also) (24), and as a reservoir for *R. maidis* and RMV. Cereal volunteers and regrowth may be important local sources of virus also (66). Less is known about the population dynamics of aphids on grasses compared with cereals, although *Metopolophium festucae*, which is often very common on grasses, is much less numerous on cereals (122). *R. padi* can be very numerous on grasses also (122), which probably explains the large size of the autumn migration as the aphids emigrate in search of *P. padus* or cereals on which to overwinter (112). A greater understanding of the population dynamics of cereal aphids on grasses would lead to more accurate forecasting of autumn infections if combined with knowledge of virus epidemiology in grasses.

Unfortunately, there are differences in the epidemiologies of the strains of BYDV on grasses and cereals (82). In Tasmania, Australia, PAV-type is more common in pasture grasses than is the RPV-type (45), whereas in Canada the reverse situation occurs (80). In the latter survey, however, a PAV-type strain was found to be more common than was the RPV-type in winter wheat and spring-sown cereals, which led to the conclusion that local virus sources were not important for disease spread in cereals. Similar results have occurred in the USA and the same conclusion has been drawn (20, 72). Although it is quite possible that this conclusion is correct for North American conditions, there is a similar inconsistency, in England, in the relative frequencies of strains in grasses and cereals as well as regional variation, too (82). The reasons for these results are unknown but probably involve the ease of acquisition by vectors of the different strains, throughout the year, and differences in aphid life cycles.

The main source of PVY and PLRV is infected potato seed tubers, although some weeds, such as *C. bursapastoris*, are hosts of PLRV and its vectors. There are also many weed species that harbor BYV and BMYV (57). Weeds are also a source of aphid vectors and are important in the spread of these viruses.

Crop Management

Rotation

The cropping sequence in a field and the combination of different crops on a farm can affect the numbers of aphid vectors and the incidence of virus.

BYDV can be particularly prevalent in cereals that follow ploughed-up grassland, as infected aphids, especially *R. padi*, are able to survive being buried for several weeks (66). These aphids then colonize the emerging cereal crops before the buried green plants die resulting in widespread infection and much damage because of the early stage at which plants are infected. Similarly, cereal volunteers and grass weeds in stubbles are colonized by migrant aphids that may be infective, before the emergence of the next cereal crop (66). If they are not destroyed, then they act as reservoirs of virus and aphid vectors. Permanent and semipermanent grassland may provide also a local source of virus and aphid vectors to neighboring cereal fields. In France (24), maize may be an important source of infective aphids in the autumn. In Britain, maize is not widely grown but as numbers of *R. padi* can reach several hundreds per plant in September and October [even though many of the winged emigrants are males or gynoparae returning to *P. padus* (50)], it may be locally important as a source of infection.

The possibility of increased virus infection in potatoes and sugar beet in the UK, comes from the large area of winter oilseed rape currently grown

(>300,000 ha). This provides not only an overwintering host for *M. persicae* but also harbors BWYV. The majority of rape plants are probably infected, although it is difficult to transmit the rape strains of BWYV to sugar beet (99).

Cultivations and Fertilizers

The effects of cultivations and fertilizers on aphid population development are usually indirect, with the primary effect being either on the host plant or on populations of natural enemies that prey on aphids. Ploughing, for example, reduces populations of soil invertebrates, including polyphagous predators, so that the use of minimum cultivations is likely to retain the highest populations of predators (31, 73). Straw burning has little direct effect on soil-inhabiting fauna but it reduces populations of such surface-living fauna (31) as spiders that are predatory on aphids (103). Fertilizers affect the physiology and nutrient status of crops and also alter the microhabitat by promoting plant growth. These factors are likely to influence aphid-population development rates and also affect colonization by aphids and natural enemies.

Little work has been done on the effects of fertilizers on the incidence of BYDV. Plumb (82) showed that increasing levels of nitrogen applied to spring barley and oats resulted in increased infection but he was not sure whether this was due to increased host susceptibility or increased numbers of aphids.

Cereal aphid densities were consistently higher (but usually not significantly so) in plots that had been ploughed compared with those that had been direct-drilled (67). Generally, this resulted in significantly more BYDV in ploughed plots. More BYDV occurred in plots that had been direct-drilled after the straw had been baled and removed rather than burned (68). Perhaps the negative effect of burning on the polyphagous predator populations was offset by the increased control of reservoirs of aphids and virus from the previous crop.

Sowing Date

Sowing date, and, hence, the time of emergence, of a crop in relation to aphid flight phenology influences the amount of virus transmission.

Experiments done at Rothamsted during the 1970s to study the effect of sowing date of winter oats, wheat, and spring barley on the level of aphid attack and BYDV incidence showed that early sowing in the autumn and late sowing in the spring increased the risk of damage (82, 83). Crops sown early in September are exposed for a longer period of time to the autumn aphid migration, which usually finishes in early November (112). Thus aphid populations are usually higher in early-sown crops (25). Similar experiments done on winter barley during the 1980s support these findings

(67, 85). For example, in the autumn of 1985, over 20% of plants sown at Rothamsted on September 13 were showing symptoms of BYDV in the spring, whereas plants sown 19 days later had less than 5% infection. In contrast, early sowing in the spring reduces the risk of damage because by the time cereal aphid migrations start in May and June the plants are well established already. Crops sown later may be infested with aphids soon after emergence, which is when damage is greatest.

A similar pattern exists for aphids and viruses on potatoes and sugar beet in the UK (15, 16). As with cereal aphids, the time of aphid flight in spring, and, hence, of subsequent development of colonies on crops, is variable, so that the relative threat to crops planted in a given location changes from year to year. If a large number of vector aphids fly soon after crop emergence, resulting problems are greater than if they fly either before emergence or late in crop growth when the resistance of mature plants (8, 96) reduces the susceptibility of the crop. Geographical location also has a major influence on both aphid and crop phenology. However, in the UK, the delay to aphid movement further north is generally greater than the delay in potato planting, and aphid peaks tend to occur at a later growth stage of the crop. Hence, crops sown early in the north usually escape infection, whereas in the south the risk is greater (17).

Variety

Varieties may be resistant to, or tolerant of, the virus or the aphid; only resistance to the aphid will be discussed here. Breeding for aphid resistance has not been done always with a view to controlling viruses. For example, although much work has been done on resistance to aphids on cereals e.g., (74, 75), little of this work is in relation to BYDV. Cereal varieties carrying the Yd_2 gene are resistant to some strains of the virus rather than to the aphids (87). In general, it would seem to be more appropriate to breed for aphid resistance when considering persistent rather than nonpersistent viruses because of the longer time required for transmission of the former. Plant characteristics that slow down aphid population development, therefore, are likely to reduce the rate of virus spread.

Wild potato, *Solanum berthaultii*, has sticky glandular trichomes on its leaves and stems that trap *M. persicae*. They are also present in hybrids of *S. berthaultii* and *Solanum tuberosum* (36). This type of resistance prevents aphids moving and reduces PLRV and PVY (44, 88). Spread of soybean mosaic virus (SMV) in soybean is reduced also by leaf pubescence that inhibits probing activity (43). There are other examples of plant resistance that inhibit probing, too, and, hence, reduce virus spread (4, 51, 70, 90). Several species of wheatgrass, *Agropyron*, show resistance to BYDV, which in most cases is due to failure of the virus to increase. In some species, however, the vectors, *R. padi* and *S. avenae*, are unable to locate the phloem, thus conferring resistance indirectly to the virus through the vector, although this varied between the two aphid species (95).

Pesticides

The chemical control strategies that can be used to prevent or reduce the amount of virus spread depend very much on the period that crops are susceptible to attack and the mode of transmission. The longer the crop is susceptible, the more applications have to be made or the more persistent the chemicals used have to be. It is also more difficult to control nonpersistent viruses as the times taken to acquire and transmit the virus are usually much shorter than the time it takes to kill the vector. Ideally, chemicals that prevent probing or repel the insect are more suitable than systemic or contact poisons.

For BYDV, chemical control consists of two approaches; (1) using herbicides to destroy the "Green Bridge" to prevent carryover of aphids and virus in a field; and (2) the use of insecticides to kill vectors to reduce the amount of secondary spread.

The use of the herbicides glyphosate or paraquat prior to ploughing reduces the incidence of BYDV in the following crop (67). However it has been shown that grassy stubbles or grassland do not harbor large amounts of virus always (66). Until these can be monitored routinely many farmers will apply herbicides prophylactically.

In situations where BYDV infection originates from viruliferous immigrant alates, the use of insecticides is concerned primarily with the reduction of secondary spread rather than the prevention of primary inoculations. Thus in England the application of synthetic pyrethroid or organophosphorus insecticides is recommended toward the end of October or early November at the end of the aphid immigration but before significant secondary spread occurs.

Potato crops are important sources of PLRV and PVY so planting high-quality seed greatly reduces the risk of secondary spread. Plants that have been infected with PLRV and PVY rarely show severe symptoms in the year of infection but planting infected tubers produces a poor crop. Thus it is particularly important to control secondary spread of these viruses in crops intended for seed. Secondary spread of both these viruses is minimal in areas where aphid vectors fly late and diseased plants can be rogued out before spread can occur.

Chemical control of secondary spread of PLRV can be achieved by using systemic insecticides, usually applied as granules at planting followed by regular (e.g., fortnightly) foliar sprays (138). However in recent years, an increasing proportion of the *M. persicae* population in the UK has become resistant to insecticides (26). If this continues an increase in PLRV incidence seems inevitable.

PVY cannot be controlled using systemic insecticides as it is acquired and inoculated too quickly. Mixtures of synthetic pyrethroids and mineral oils provide some control (37) but repeated use can cause a resurgence of resistant *M. persicae* at the end of the season (33). This resurgence does not result in higher tuber infection (53), probably because it occurs at such a late stage in crop growth that mature plant resistance (8) prevents spread of virus to the tubers. Two early-season applications of pyrethroid–oil mixtures are adequate for control of PVY (52).

Systemic insecticides are used to control virus spread in sugar beet but again, there are problems regarding insecticide resistance in the major vector, *M. persicae*. Recent increases in insecticide resistance probably explain why the current preseason virus-warning scheme (55) that is based on data collected by the Rothamsted Insect Survey since 1965 tends to underestimate virus incidence. The British Sugar company employs fieldpersons whose duties include surveying crops for the presence of aphids. On the basis of these counts, spray warnings are issued on a factory area basis. Such warnings lead to the more efficient control of virus spread as they reduce the need for prophylactic insecticide usage and slow down the spread of insecticide resistance. A greater understanding of the relationship between aerial and crop populations could reduce the time needed for surveying crops and result in a more efficient warning scheme.

Summary

1. The effects of weather, biotic factors, and crop management on aphid biology and behavior and the consequences for the epidemiology of persistent, semipersistent, and nonpersistent viruses are reviewed.
2. Overwintering survival of aphids, and, hence, subsequent virus spread are dependent on the interaction between weather and life-cycle strategy. A higher proportion of aphids overwintering as eggs survive after a cold winter is compared with aphids overwintering viviparously. In mild winters, aphids overwintering viviparously increase in number, whereas the numbers of eggs cannot increase.
3. Similarly, the effect of spring population development on virus spread is related more to the aphid life cycle than to the type of virus transmitted. Populations are likely to develop more rapidly from anholocyclic clones during mild springs than from clones that overwinter as eggs. Anholocyclic clones and holocyclic monoecious clones are more likely to acquire viruses of agricultural importance from overwintering host plants than from holocyclic heteroecious clones that overwinter as eggs, usually on woody hosts. Temperature is the main weather factor influencing spring population development.
4. Persistent viruses are likely to be dispersed over larger areas than semi- and nonpersistent viruses by the longer-term nature of their association with aphids. However, most virus spread seems to occur over relatively short distances. Windspeed and direction, together with

temperature, are the most important weather factors influencing aphid, and, hence, virus dispersal.

5. Secondary spread of persistent viruses is most likely to be caused by apterous aphids, whereas alate aphids are probably of increased importance in the spread of nonpersistent viruses. The weather factors influencing secondary spread are largely unquantified but it is likely that temperature is important.

6. Many such intrinsic aphid effects as clonal variation in relation to proportions of viruliferous aphids depend more on aphid biology than on the type of virus being transmitted.

7. The effects of aphid natural enemies on virus spread are largely unknown but they are likely to be more significant with persistent viruses because of the time taken to affect aphid population development. But, through colony disturbance, they may result in increased aphid movement and, hence, greater virus spread.

8. Alternative aphid host plants affect virus spread, for the most part independently of the type of virus and are particularly important during the time crops are not present above ground.

9. Management practices, such as rotations, cultivations, and fertilizers, are likely to have similar effects for the different types of virus.

10. The effects of sowing date on virus incidence are influenced more by aphid flight phenology than by the type of virus.

11. The development of varieties resistant to aphids is likely to be more effective against persistent than nonpersistent viruses because of the longer time it takes to acquire and inoculate the former. Varieties that deter probing or movement, however, could be useful in control of nonpersistent viruses.

12. Similarly, with pesticides, to be effective against nonpersistent viruses they need to be fast-acting to prevent probing. Antifeedants may be useful in this respect. Pesticides are effective at reducing the secondary spread of persistent viruses.

13. To improve the forecasting and control of aphid-borne viruses, it is essential to gain a greater understanding of the interactions between the vectors and abiotic, biotic, and crop management factors in relation to virus epidemiology.

References

1. A'Brook, J., 1981, Some observations in west Wales on the relationships between numbers of alate aphids and weather, *Ann. Appl. Biol.* **97**:11–15.

2. Adlerz, W.C., 1974, Wind effects on spread of watermelon mosaic virus. I. From local virus sources to watermelons, *J. Econ. Entomol.* **67**:361–364.

3. Ajayi, O. and Dewar, A.M., 1982, The effect of barley yellow dwarf virus on honeydew production by the cereal aphids *Sitobion avenae* and *Metopolophium dirhodum. Ann. Appl. Biol.* **100**:203–12.

4. Atiri, G.I., Ekpo, E.J.A., and Thottapilly, G., 1984, The effect of aphid-resistance in cowpea on infestation and development of *Aphis craccivora* and the transmission of cowpea aphid-borne mosaic virus, *Ann. Appl. Biol.* **104**:339–346.

5. Ankersmit, G.W. and Dijkman, H., 1983, Alatae production in the cereal aphid *Sitobion avenae*, *Neth. J. Pl. Path.* **89**:105–112.

6. Bale, J.S., Harrington, R., and Clough, M.S., 1988, Low temperature mortality of the peach-potato aphid, *Myzus persicae, Ecol. Entomol.* **13**:121–129.

7. Barker, H., 1989, Specificity of the effect of sap-transmissible viruses in increasing the accumulation of Luteoviruses in co-infested plants, *Ann. Appl. Biol.* **115**:71–78

8. Beemster, A.B.R., 1972, Virus translocation in potato plants and mature-plant resistance, de Bokx, J.A. (ed): in Virus of Potatoes and Seed Potato Production, Pudoc, Wageningen, pp. 144–151.

9. Bishop, A.L., Anderson, J.M.E., and Hales, D.F., 1986, Predators: agents for biological control, (ed): McLean, G.D., Garrett, R.G., and Ruesink, W.G. in Plant Virus Epidemics Monitoring, Modelling and Predicting Outbreaks, Academic Press, Sydney, pp. 75–94.

10. Boakye, D.B. and Randles, J.W., 1974, Epidemiology of lettuce necrotic yellows virus in South Australia. III. Virus transmission parameters and vector feeding behaviour on host and non-host plants, *Aust. J. Agric. Res.* **25**:791–802.

11. de Bokx, J.A., 1981, Potato virus Y. In *CMI/AAB Descriptions of Plant Viruses*. 242. Commonwealth Agricultural Bureaux/Association of Applied Biologists.

12. Broadbent, L., 1948, Factors affecting aphid flight. 8th *Int. Congr. Entomol.*, Stockholm, pp. 619–21.

13. Broadbent, L., 1949, Factors affecting the activity of alatae of the aphids *Myzus persicae* (Sulzer) and *Brevicoryne brassicae* (L.), *Ann. Appl. Biol.* **36**:40–62.

14. Broadbent, L., Cornford, C.E., Hull, R., and Tinsley, T.W., 1949, Overwintering of aphids, especially *Myzus persicae* (Sulzer), in root clamps, *Ann. Appl. Biol.* **36**:513–524.

15. Broadbent, L., Gregory, P.H., and Tinsley, T.W., 1952, The influence of planting date and manuring on the incidence of virus diseases in potato crops, *Ann. Appl. Biol.* **39**:509–524.

16. Broadbent, L., Heathcote, G.D., McDermott, N., and Taylor, C.E., 1957, The effect of date of planting and of harvesting potatoes on virus infection and on yield. *Ann. Appl. Biol.* **45**:603–622.

17. Cadman, C.H. and Chambers, J., 1960, Factors affecting the spread of aphid-borne viruses in potato in Eastern Scotland. III. Effect of planting date, roguing and age of crop on the spread of potato leaf-roll and Y viruses. *Ann. Appl. Biol.* **48**:729–738.

18. Carter, N., 1989, The role of natural enemies in arable crops, *Proc. Symposium: Insect Control Strategies and the Environment*, 1989, 51–69.

19. Carter, N., Mclean, I.F.G., Watt, A.D., and Dixon, A.F.G., 1980, Cereal aphids: a case study and review, Coaker, T.H. (ed): in Applied Biology, Academic Press, London, **5**:272–348.

20. Clement, D.L., Lister, R.M., and Foster, J.E., 1986, ELISA-based studies

on the ecology and epidemiology of barley yellow dwarf virus, *Phytopathology* **76**:86–92.

21. Cockbain, A.J., Gibbs, A.J., and Heathcote, G.D., 1963, Some factors affecting the transmission of sugar-beet mosaic virus and pea mosaic viruses by *Aphis fabae* and *Myzus persicae*, *Ann. Appl. Biol.* **52**:133–143.

22. Davies, W.M., 1936, Studies on aphids infesting the potato crop. V. Laboratory experiments on the effect of wind velocity on the flight of *Myzus persicae* Sulz, *Ann. Appl. Biol.* **23**:401–408.

23. Dean, G.J., 1974, Effects of temperature on the cereal aphids, *Metopolophium dirhodum* (Wlk.), *Rhopalosiphum padi* (L.) and *Macrosiphum avenae* (F.) (Hem., Aphididae), *Bull. Ent. Res.* **63**:401–409.

24. Dedryver, C.A. and Robert, Y., 1981, Ecological role of maize and cereal volunteers as reservoirs for gramineae virus transmitting aphids. *Proc. 3rd Conf. Virus Diseases of Gramineae in Europe, 1980*, Plumb, R.T. (ed): pp. 61–66.

25. Dewar, A.M. and Carter, N., 1984, Decision trees to assess the risk of cereal aphid (Hemiptera: Aphididae) outbreaks in summer in England, *Bull. Ent. Res.* **74**:387–398.

26. Dewar, A.M., Devonshire, A., and ffrench-Constant, R., 1988, The rise and rise of the resistant aphid, *British Sugar Beet Rev.* **56**:40–43.

27. Dixon, A.F.G., 1971, The life-cycle and host preferences of the bird cherry-oat aphid, *Rhopalosiphum padi* L., and their bearing on the theories of host alternation in aphids, *Ann. Appl. Biol.* **68**:135–147.

28. Dry, W.W. and Taylor, L.R., 1970, Light and temperature thresholds for take-off by aphids, *J. Animal Ecol.* **39**:493–504.

29. Duffus, J.E., 1963, Possible multiplication in the aphid vector of sowthistle yellow vein virus, a virus with an extremely long insect latent period, *Virology* **21**:194–202.

30. Dunning, R.A., Baker, A.N., Windley, and R.F., 1975, Carabids in sugar beet crops and their possible role as aphid predators, *Ann. Appl. Biol.* **10**:125–128.

31. Edwards, C.A., 1977, Investigations into the influence of agricultural practice on invertebrates, *Ann. Appl. Biol.* **87**:515–520.

32. Elliott, W.M. and Kemp, W.G., 1979, Flight activity of the green peach aphid (Homoptera: Aphididae) during the vegetable growing season at Harrow and Jordon, Ontario, *Proc. Ent. Soc. Ontario* **110**:19–28.

33. ffrench-Constant, R.H., Harrington, R., and Devonshire, A.L., 1988, Effect of repeated applications of insecticides to potatoes on numbers of *Myzus persicae* (Sulzer) (Hemiptera: Aphididae) and on the frequencies of insecticide-resistant variants, *Crop. Prot.* **7**:55–61.

34. Fisken, A.G., 1959, Factors affecting the spread of aphid-borne viruses in potato in Eastern Scotland. I. Overwintering of potato aphids, particularly *Myzus persicae* (Sulzer) in England, *Ann. Appl. Biol.* **47**:264–273.

35. Gabriel, W., 1965, The influence of temperature on the spread of aphid-borne potato virus diseases, *Ann. Appl. Biol.* **56**:461–475.

36. Gibson, R.W., 1974, Aphid-trapping glandular hairs on hybrids of *Solanum tuberosum* and *S. berthaullii*, *Potato Res* **17**:152–154.

37. Gibson, R.W. and Cayley, G.R., 1984, Improved control of potato virus Y by mineral oil plus the pyrethroid cypermethrin applied electrostatically, *Crop.*

Prot. **3**:469–478.

38. Gilbert, N., 1980, Comparative dynamics of a single-host aphid. I. The evidence, *J. Anim. Ecol.* **49**:351–369.

39. Gildow, F.E., 1980, Increased production of alatae by aphids reared on oats infected with barley yellow dwarf virus. *Ann. Entomol. Soc. Am.* **73**:343–347.

40. Gildow, F.E., 1983, Influence of barley yellow dwarf virus-infected oats and barley on morphology of aphid vectors, *Phytopathology* **73**:1196–1199.

41. Govier, D.A. and Kassanis, B., 1974, Evidence that a component other than the virus particle is needed for aphid transmission of potato virus Y, *Virology* **57**:285–286.

42. Govier, D.A. and Kassanis, B., 1974, A virus-induced component of plant sap needed when aphids acquire potato virus Y from purified preparations, *virology* **61**:420–426.

43. Gunasinghe, U.B., Irwin, M.E., and Kampmeier, G.E., 1988, Soybean leaf pubescence affects aphid vector transmission and field spread of soybean mosaic virus, *Ann. Appl. Biol.* **112**:259–272.

44. Gunenc, Y. and Gibson, R.W., 1980, Effect of glandular foliar hairs on the spread of potato virus Y. *Potato Res* **23**:345–351.

45. Guy, P.L., Johnstone, G.R. and Duffus, J.E., 1986, Occurrence and identity of barley yellow dwarf viruses in Tasmanian pasture grasses, *Aust. J. Agric. Res.* **37**:43–53.

46. Haine, E., 1955, Aphid take-off in controlled wind speeds, *Nature*, London, **175**:474–475.

47. Halbert, S.E. and Pike, K.S., 1985, Spread of barley yellow dwarf virus and relative importance of local aphid vectors in central Washington, *Ann. Appl. Biol.* **107**:387–395.

48. Halgren, L.A. and Taylor, L.R., 1968, Factors affecting flight responses of aliencolae of *Aphis fabae* Scop. and *Schizaphis graminum* Rondani (Homoptera: Aphididae). *J. Animal Ecol.* **37**:583–593.

49. Hand, S.C., 1980, Overwintering of cereal aphids, *IOBC WPRS Bull.* **3**:59–61.

50. Hand, S.C., Carrillo, J.R., 1982, Cereal aphids on maize in southern England, *Ann. Appl. Biol.* **100**:39–47.

51. Haniotakis, G.E. and Lange, W.H., 1973, Beet yellows virus resistance in sugar beets: mechanism of resistance, *J. Econ. Entomol.* **67**:25–28.

52. Harrington, R., 1988, Prospects for controlling the spread of potato virus Y and for using aphid data to assess virus spread, *The Seed Potato* **27**:6.

53. Harrington, R., Bartlet, E., Riley, D.K., Clark, S.J., and ffrench-Constant, R.H., 1989, Resurgence of insecticide-resistant *Myzus persicae* on potatoes treated repeatedly with cypermethrin and mineral oil, *Crop Prot.* **8**:340–348.

54. Harrington, R. and Cheng, X-N., 1984, Winter mortality, development and reproduction in a field population of *Myzus persicae* (Sulzer) (Hemiptera: Aphididae) in England, *Bull. Ent. Res.* **74**:633–640.

55. Harrington, R., Dewar, A.M. and George, B., 1989, Forecasting the incidence of virus yellows in sugar beet in England, *Ann. Appl. Biol.* **114**:459–469.

56. Heathcote, G.D. and Cockbain, A.J., 1966, Aphids from mangold clamps and their importance as vectors of beet viruses, *Ann. Appl. Biol.* **57**:321–336.

57. Heathcote, G.D., Dunning, R.A. and Wolfe, M.D., 1965, Aphids on sugar beet and some weeds in England, and notes on weeds as a source of beet viruses, *Pl. Path.* **14**:1–10.
58. Hille Ris Lambers, D., 1972, Aphids: their life cycles and their role as virus vectors, (ed): de Bokx, J.A. in Viruses of Potatoes and Seed Potato Production, Pudoc, Wageningen, pp. 36–64.
59. Hodson, A.C. and Cook, E.F., 1960, Long-range aerial transport of the harlequin bug and the greenbug into Minnesota, *J. Econ. Entomol.* **53**:604–608.
60. Howell, P.J., 1977, Recent trends in the incidence of aphid-borne viruses in Scotland. *Proc. Symposium Problems of Pest and Disease Control in Northern Britain, Dundee, 1977*, pp. 26–28.
61. Inayatullah, C., Fargo, W.S. and Webster, J.A., 1987, Use of multivariate models in differentiating greenbug (Homoptera: Aphididae) biotypes and morphs, *Environ. Entomol.* **16**:839–846.
62. Jepson, P.C. and Green, R.E., 1982, Prospects for improving control strategies for sugar-beet pests in England, Coaker, T.H. (ed): in *Advances in Applied Biology*, Academic Press, London, **7**:175–250.
63. Johnson, C.G., 1969, Migration and Dispersal of Insects by Flight, Methuen, London, 763 p.
64. Jones, F.G.W. and Dunning, R.A., 1972, Sugar beet pests, MAFF Bulletin 162, HMSO, London, 114 p.
65. Kemp, W.G. and Troup, P.A., 1978, A weather index to forecast potential incidence of aphid-transmitted virus diseases of peppers in the Niagaru peninsula, *Can. J. Plant Sci.* **58**:1025–1028.
66. Kendall, D.A., 1986, Volunteer cereals, grass weeds and swards as sources of cereal aphids and barley yellow dwarf virus (BYDV), *Proc. EWRS Symposium 1986, Economic Weed Control*, pp. 201–208.
67. Kendall, D.A., Chinn, N.E., Wiltshire, C.W., and Bassett, P. 1988. Effects of crop agronomy and agrochemical use on the incidence of barley yellow dwarf virus in autumn sown cereals, *Aspects Appl. Biol.* **17**:143–151.
68. Kendall, D.A., Smith, B.D., Chinn, N.E., and Wiltshire, C.W., 1986, Cultivation, straw disposal and barley yellow dwarf virus infection in winter cereals, *1986 British Crop Protection Conference—Pests and Diseases*, **3**:981–987.
69. Lamb, R.J. and MacKay, P.A., 1979, Variability in migratory tendency within and among natural populations of the pea aphid, *Acyrthosiphon pisum*, *Oecologia* **39**:289–299.
70. Lapointe, S.L. and Tingey, W.M., 1984, Feeding response of the green peach aphid (Homoptera: Aphididae) to potato glandular Trichomes, *J. Econ. Entomol.* **77**:386–389.
71. Lim, W.L. and Hagedorn, D.J., 1977, Bimodal transmission of plant viruses, Harris, K.F., Maramorosch, K. (ed): in Aphids as Virus Vectors, Academic Press, New York, pp. 237–251.
72. Lister, R.M., Clement, D., Skaria, M. and Foster, J.E., 1984, Biological differences between barley yellow dwarf viruses in relation to their epidemiology and host reactions, *Proc. BYD Workshop, 1983*, pp. 16–25.
73. Luff, M.L., 1987, Biology of polyphagous ground beetles in agriculture, *Agric. Zoo. Rev.* **2**:237–278.

74. Lowe, H.J.B., 1978, Detection of resistance to aphids in cereals, *Ann. Appl. Biol.* **88**:401–406.
75. Lowe, H.J.B., 1980, Resistance to aphids in immature wheat and barley, *Ann. Appl. Biol.* **95**:129–135.
76. Mackay, P.A., 1989, Clonal variation in sexual morph production in *Acyrthosiphon pisum* (Homoptera: Aphididae). *Environ. Entomol.* **18**:558–562.
77. Medler, J.T. and Smith, P.W., 1960, Greenbug dispersal and distribution of barley yellow dwarf virus in Wisconsin, *J. Econ. Entomol.* **53**:473–474.
78. Miller, P.W., 1952, Relation of temperature to persistence of strawberry yellows virus complex in the strawberry aphid, *Phytopathology* **42**:517.
79. Miller, J.W. and Coon, B.F., 1964, The effect of barley yellow dwarf virus on the biology of its vector the English grain aphid, *Macrosiphum granarium. J. Econ. Entomol.* **57**:970–974.
80. Paliwal, Y.C., 1982, Role of perennial grasses, winter wheat, and aphid vectors in the disease cycle and epidemiology of barley yellow dwarf virus. *Can. J. Plant Pathol.* **4**:367–374.
81. Plumb, R.T., 1976, Barley yellow dwarf virus in aphids caught in suction traps 1969–73, *Ann. Appl. Biol.* **83**:53–59.
82. Plumb, R.T., 1977, Grass as a reservoir of cereal viruses, *Ann. Phytopathol.* **9**:361–364.
83. Plumb, R.T., 1981, Chemicals in the control of cereal virus diseases. Jenkyn, J.F. and Plumb, R.T. (ed): in Strategies for the Control of Cereal Disease, Blackwell, Oxford, pp. 135–145.
84. Plumb, R.T., 1983, Barley yellow dwarf virus—a global problem. Plumb, R.T. and Thresh, J.M. (ed): in Plant Virus Epidemiology, Blackwell, Oxford, pp. 185–198.
85. Plumb, R.T., 1988, Opportunities for the integrated control of barley yellow dwarf viruses in UK. *Aspects Appl. Biol.* **17**:153–161.
86. Randles, J.W. and Crowley, N.C., 1970, Epidemiology of lettuce necrotic yellows virus in South Australia. I. Relationship between disease incidence and activity of *Hyperomyzus lactucae* (L.). *Aust. J. Agric. Res.* **21**:447–453.
87. Rasmussan, D.C. and Schaller, C.W., 1959, The inheritance of resistance in barley to the barley yellow dwarf virus. *Agron. J.* **51**:661–664.
88. Rizvi, S.A.H. and Raman, K.V., 1983, Glandular hairs in *Solanum berthaultii* Hawkes defend plants against several insect and mite pests by physically trapping and immobilizing foliage-feeding pests, Hooker, W.J. (ed): in Research for the potato in the year 2000, International Potato Centre, Peru, pp. 162–163.
89. Robert, Y. and Rouze-Jouan, J., 1978, Recherches ecologiques sur les pucerons *Aulacorthum solani* KLTB, *Macrosiphum euphorbiae* Thomas et *Myzus persicae* Sulz. dans l'Quest de la France. I. Etude de l'activite de vol de 1967 a 1976 en culture de pomme de terre, *Ann. Zool. Ecol. Anim.* **10**:171–185.
90. Roberts, J.J. and Foster, J.E., 1983, Effect of leaf pubescence in wheat on the bird cherry oat aphid (Homoptera: Aphidae), *J. Econ. Entomol.* **76**:1320–1322.
91. Rochow, W.F., 1970, Barley yellow dwarf virus. *CMI/AAB Descriptions of Plant Viruses* **32**.

92. Rochow, W.F. and Eastop, V.F., 1966, Variation within *Rhopalosiphum padi* and transmission of barley yellow dwarf virus by clones of four aphid species, *Virology* **30**:286–296.
93. Roitberg, B.D. and Myers, J.H., 1978, Effect of adult coccinellidae on the spread of a plant virus by an aphid, *J. Appl. Ecol.* **15**:775–779.
94. Russell, G.E., 1963, Some factors affecting the relative incidence, distribution and importance of beet yellows and sugar beet mild yellowing virus in Eastern England, 1955–62, *Ann. Appl. Biol.* **52**:405–413.
95. Shukle, R.H., Lampe, D.J., Lister, R.M., and Foster, J.E., 1987, Aphid feeding behaviour: relationship to barley yellow dwarf virus resistance in *Agropyron* species, *Phytopathology* **77**:725–729.
96. Sigvald, R., 1984, The relative efficiency of some aphids as vectors of potato virus Y° (PVY°), *Potato Res.* **27**:289–290.
97. Simons, J.N., 1966, Effects of temperature and length of acquisition feeding time on transmission of nonpersistent viruses by aphids, *J. Econ. Entomol.* **59**:1056–1062.
98. Singh, A.B., 1972, Studies on the transmission of papaya mosaic virus by *Aphis gossypii* Glover, *In. J. Entomol.* **34**:240–245.
99. Smith, H.G. and Hinckes, J.A., 1985, Studies on beet western yellows virus in oilseed rape (*Brassica napus* ssp. *oleifera*) and sugar beet (*Beta vulgaris*), *Ann. Appl. Biol.* **107**:473–484.
100. Smith, P.R. and Plumb, R.T., 1981, Barley yellow dwarf virus infectivity of cereal aphids trapped at two sites in Australia, *Aust. J. Agric. Res.* **32**:249–255.
101. Smith, R.K., 1981, Studies on the ecology of cereal aphids and prospects for integrated control. *PhD Thesis*, London, University of London, 400 pp.
102. Sopp. P.I. and Chiverton, P.A., 1987, Autumn predation of cereal aphids by polyphagous predators in southern England: a 'first look' using ELISA, *IOBC WPRS Bull.* **10**:103–108.
103. Sunderland, K.D., Fraser, A.M., and Dixon, A.F.G., 1986, Field and laboratory studies on money spiders (Linyphiidae) as predators of cereal aphids, *J. Appl. Ecol.* **23**:433–447.
104. Syller, J., 1987, The influence of temperature on the transmission of potato leafroll virus by *Myzus persicae* Sulz, *Potato Res.* **30**:47–58.
105. Sylvester, E.S., 1949, Beet-mosaic virus—green peach aphid relationships, *Phytopathology* **39**:417–424.
106. Sylvester, E.S., 1964, Some effects of temperature on the transmission of cabbage mosaic virus by *Myzus persicae*, *J. Econ. Entomol.* **57**:538–544.
107. Sylvester, E.S., 1973, Reduction of excretion, reproduction and survival in *Hyperomyzus lactucae* fed on plants infected with isolates of sowthistle yellow vein virus, *Virology* **56**:632–635.
108. Sylvester, E.S. and Richardson, J., 1966, Some effects of temperature on the transmission of pea enation mosaic virus and on the biology of the pea aphid vector, *J. Econ. Entomol.* **59**:255–261.
109. Takada, H., 1988, Interclonal variation in the photoperiodic response for sexual morph production in Japanese *Aphis gossypii* Glover (Hom., aphididae), *J. Appl. Ent.* **106**:188–197.
110. Tamada, T. and Harrison, B.D., 1981, Quantitative studies on the uptake and retention of potato leafroll virus by aphids in laboratory and field conditions, *Ann. Appl. Biol.* **98**:261–276.

111. Tanaka, S. and Shiota, H., 1970, Latent period of potato leafroll virus in the green peach aphid (*Myzus persicae* Sulz.), *Ann. Phytopath. Soc. Japan* **36**:106–111.

112. Tatchell, G.M., Plumb, R.T., and Carter, N., 1988, Migration of alate morphs of the bird cherry aphid (*Rhopalosiphum padi*) and implications for the epidemiology of barley yellow dwarf virus, *Ann. Appl. Biol.* **112**:1–11.

113. Taylor, L.R., 1963, Analysis of the effect of temperature on insects in flight, *J. Anim. Ecol.* **32**:99–117.

114. Taylor, L.R., 1977, Migration and the spatial dynamics of an aphid, *Myzus persicae, J. Anim. Ecol.* **46**:411–423.

115. Taylor, L.R., 1986, Synoptic dynamics, migration and the Rothamsted Insect Survey, *J. Anim. Ecol.* **55**:1–38.

116. Taylor, L.R., 1986, The distribution of virus disease and the migrant vector aphid. McLean, G.D., Garrett, R.G., and Ruesink, W.G. (ed): in *Plant Virus Epidemics Monitoring, Modelling and Predicting Outbreaks*, Academic Press, Sydney, pp. 35–57.

117. Taylor, L.R., French, R.A., Woiwod, I.P., Dupuch, M., and Nicklen, J., 1981, Synoptic monitoring for migrant insect pests in Great Britain and Western Europe. I. Establishing expected values for species content, population stability and phenology of aphids and moths, *Rothamsted Expt. Station. Rep. 1980*, Part 2, 41–104.

118. Turl, L.A.D., 1977, Aphid epidemiology in virus control. *Proc. Symposium Problems of Pest and Disease Control in Northern Britain, Dundee, 1977*, pp. 31–32.

119. Turl, L.A.D., 1980, An approach to forecasting the incidence of potato and cereal aphids in Scotland, *EPPO Bull.* **10**:135–141.

120. Turl, L.A.D., 1983, The effect of winter weather on the survival of aphid populations on weeds in Scotland, *EPPO Bull.* **13**:139–143.

121. van der Broek, L.J. and Gill, C.C., 1980, The median latent periods for three isolates of barley yellow dwarf virus in aphid vectors, *Phytopathology* **70**:644–646.

122. Vickerman, G.P., 1982, Distribution and abundance of cereal aphid parasitoids (*Aphidius* spp.) on grassland and winter wheat. *Ann. Appl. Biol.* **101**:185–190.

123. Wallin, J.R. and Loonan, D.V., 1971, Low level jet winds, aphid vectors, local weather and barley yellow dwarf virus outbreaks, *Phytopathology* **61**:1068–1070.

124. Wallin, J.R., Peters, D., and Johnson, L.C., 1967, Low-level jet winds, early cereal aphid and barley yellow dwarf detection in Iowa, *Plant Disease Reporter* **51**:527–530.

125. Walters, K.F.A. and Dewar, A.M., 1986, Overwintering strategy and the timing of the spring migration of the cereal aphids *Sitobion avenae* and *Sitobion fragariae, J. Appl. Ecol.* **23**:905–915.

126. Walters, K.F.A. and Dixon, A.F.G., 1984, The effect of temperature and wind on the flight activity of cereal aphids, *Ann. Appl. Biol.* **104**:17–26.

127. Watson, M.A., 1936, XII-Factors affecting the amount of infection obtained by Aphis transmission of the virus Hy. III, *Phil. Trans. Royal Soc. Lond. Ser. B*, **226**:457–489.

128. Watson, M.A., 1938, Further studies on the relationships between *Hyos-*

cyomus virus 3 and the aphis *Myzus persicae* (Sulz.) with special reference to the effects of fasting, *Proc. R. Soc. Lond., Ser. B*, **125**:144–170.

129. Watson, M.A., 1946, The transmission of beet mosaic and beet yellows viruses by aphids; a comparative study of a non-persistent and persistent virus having host plants and vectors in common, *Proc. R. Soc. Lond., Ser. B*, **133**:200–219.

130. Watson, M.A., Heathcote, G.D., Lauckner, F.B., and Sowray, P.A., 1975, The use of weather data and counts of aphids in the field to predict the incidence of yellowing viruses of sugar beet crops in England in relation to the use of insecticides, *Ann. Appl. Biol.* **81**:181–198.

131. Watson, M.A., Roberts, F.M., 1939, A comparative study of the transmission of *Hyoscyamus* virus 3, potato virus Y and cucumber virus 1 by the vectors *Myzus persicae* (Sulz.), *M. circumflexus* (Buckton), and *Macrosophum gei* (Koch), *Proc. R. Soc. Lond., Ser. B*, **127**:543–576.

132. Watt, A.D. and Dixon, A.F.G., 1981, The role of cereal growth stages and crowding in the induction of alatae in *Sitobion avenae* and its consequences for population growth, *Ecol. Entomol.* **6**:441–447.

133. Webster, J.A., and Inayatullah, C. 1985, Aphid biotypes in relation to host plant resistance: a selected bibliography. *Southwest. Entomol.* **10**:116–125.

134. Weidemann, H.L., 1981, Die Einstichsaktivität von Blattlausen, gemessen an der Anzahl von Speichelscheiden, im Hinblick auf Virusübertragungen, *Z. Ang. Ent.* **92**:92–98.

135. Wellington, W.G., 1945, Conditions governing the distribution of insects in the free atmosphere. IV. Distributive processes of economic importance, *Can. Entomol.* **77**:69–74.

136. Wiktelius, S., 1977, The importance of southerly winds and other weather data on the incidence of sugar beet yellowing viruses in southern Sweden, *Swedish J. Agric. Res.* **7**:89–95.

137. Williams, C.B., 1940, An analysis of four years' captures of insects in a light trap. II. The effect of weather conditions in insect activity; and the estimation and forecasting of changes in insect activity; and the estimation of changes in the insect population, *Trans. R. Ent. Soc. Lond.* **90**:79–131.

138. Woodford, J.A.T., Harrison, B.D., Aveyard, C.S., and Gordon, S.C., 1983, Insecticidal control of aphids and the spread of potato leafroll virus in potato crops in Eastern Scotland, *Ann. Appl. Biol.* **103**:117–130.

139. Wratten, S.D. 1987, The effectiveness of natural enemies. Burn, A.J., Coaker, T.H. and Jepson, P.C. (ed): in *Integrated Pest Management*, Academic Press, London, pp. 89–112.

140. Wratten, S.D. and Pearson, J., 1982, Predation of sugar beet aphids in New Zealand, *Ann. Appl. Biol.* **101**:178–181.

141. Zeyen, R.J., Stromberg, E.L. and Kuehnast, E.L. 1987, Long-range transport hypothesis for maize dwarf mosaic virus: history and distribution in Minnesota, USA, *Ann. Appl. Biol.* **111**:325–336.

3
Cyclic Epidemics of Aphid-Borne Potato Viruses in Northern Seed-Potato-Growing Areas

Richard H. Bagnall

Introduction

In the "potato belt," along the northern New Brunswick–Maine border, the author sees epidemiology of aphid-transmitted viruses in seed potatoes as a clash between two cycles (7, 8). The leading participants in the act are the "native" northern aphids, potential vectors of the mosaic viruses, potato viruses Y (PVY) and A (PVA), and their more famous "southern cousin," *Myzus persicae* Sulz., the principal vector of the potato leaf roll virus (PLRV).

Much of this chapter is an outline of papers in a current series on periodic epidemics; but I have reread some of the original citations and have encountered others that are of interest. There are also some new developments from the field. The potato leaf roll cycle and the ubiquitous mosaic diseases can be traced back into the 18th century, so it is necessary to draw upon Britain and Europe for historical material. I concentrate largely on the North American events since the early part of this century; but, again, it is useful to consider some European work, especially with respect to vectors of PVY. At the present time, North American workers in this field are considerably outnumbered.

There is good circumstantial evidence that the incitants of both diseases, mosaic and leaf roll, go back to the earliest times of potato culture in both North America and Europe.

The Curl in Potatoes—1750 to 1905

Cause of the Curl

Early descriptions of "the curl," a group of degenerative-type diseases known in Britain and western Europe during the 18th and 19th centuries,

Richard H. Bagnall, Agriculture Canada Research Station, Fredericton, New Brunswick, Canada E3B-4Z7.
© 1991 Springer-Verlag New York, Inc. *Advances in Disease Vector Research*, Volume 7.

clearly depict participation of PVY. The possible presence of PLRV was challenged by Atanasoff (6) who alleged that the curl was simply his own "stipple-streak" disease (5). But Davidson (14) suggested that Atanasoff was obsessed with stipple-streak, now known as a manifestation of PVY, and saw it even where descriptions applied more aptly to other diseases, some caused by bacteria and fungi. Davidson cited instances where leaf roll had clearly been described. We now know that PLRV occurs widely in wild Solanum species (31) and in the Andigena (40) in South America, the original sources of our commercial potatoes. Moreover, when leaf roll was first identified as a distinct disease, it was already widely distributed in both Europe and North America. And Quanjer stated that when he commenced work about 1907, only the term *curl* was then being used in Holland. He later learned to distinguish leaf roll and mosaic, but observed that most truly curled or degenerate plants were afflicted with both (38). I am proceeding on the premise that both PLRV and PVY—probably both in a number of different strains—were actually part of the curl complex.

Background on the Curl

I side with Davidson (14) in the PLRV argument, on the basis of circumstantial evidence. But I draw liberally from both his and Atanasoffs' (6) reviews, and also from others by Orton (34), Pethybridge (35, 36), Quanjer (37), and Salaman (43, 44, 45). They have sifted through a truly voluminous literature, much of it, in keeping with the earlier times— dogmatic and uninformative. This covers a period from 1750 to 1905.

The first major outbreak of the disease known as the curl in Britain and in Germany as *Krauselkrankheit*, about 1775–1784, was relatively well documented at that time. Those close to the seed potato industry today would note a familiar ring about the knowledge or practices advanced in the late 18th century (6, 14): for example, that plants raised from true seed were initially healthy, but sooner or later, succumbed to the curl; of roguing, and growing a seed plot some distance from other potato crops; that seed lots grown from healthy stock had less curl when lifted early than did those left till maturity; the testing of seed lots during the winter in a hot bed; that certain varieties suffered less from the curl than did others; and the widespread use of seed potatoes from high altitude or northern areas where curl was absent or scarce. These control techniques were little improved upon until the 1970s.

Anderson (Refs. in 6) suggested that the curl disease arose from "infected seed" (this meant *inherited*, from his description), but that it was possibly communicated by "juxtaposition" (closeness or in contact). This was 1778, however, so it was one or the other. The distinction between true inheritance and a disease principle being passed on through the tubers, as well as the idea of an infectious principle being transmitted from plant to plant, had yet to register. But with Anderson speaking of the need for further research, we wonder that the concept, with respect to the curl, took another 138 years to develop, as we will see later.

Partly, there seemed to be no need of research, for the matter of the cause had been virtually settled (6, 14). And the forthright statements to this effect were in sharp contrast to the cautious modesty with which the observations noted above were put forward. Some dissenting words were written, but prizes and medals were awarded by scientific and agricultural societies for the knowledge that the curl was degeneration due to long continued asexual reproduction. Moreover, the success of early lifting was widely ascribed to the use of immature seed. And the therapeutic effect of growing seed lots at high altitude was still being advanced in the 1920s. Some of the accurate observations were subjected to different interpretations.

The Lifespan of Varieties and Epidemics of the Curl

The lifespan of a potato variety before it degenerated was discussed frequently in the literature. This ranged from 3 to 14 years in England, 20 to 30 in Scotland and 30 or more in Ireland (6, 14). This lifespan became the subject of comment by breeders in the latter part of the 19th century (15, 44). These estimates bore a resemblance to the length of the periods when the curl was known to be scarce or absent. In that context, some writers commented that the curl was seen only occasionally prior to 1775, and was again scarce after 1784 through the 1790s, from 1812 to the 1840s, a period of years before 1875, and between 1880 and 1905.

Meanwhile, there were serious outbreaks, **1775–1784** (6, 14), and **1875–1879** (6, 7). There have been persistent references to a single, continuous, outbreak from the 1770s to 1812, but I have concluded that there were actually two, the second just prior to **1812**, as we have evidence of a decline in the 1790s. There is evidence for a further outbreak in the **1840s** from both Britain and the United States. Several writers described the curl as being common about 1850. Moreover, breeders at the time, such as Patterson of Britain (15) and Goodrich of the United States (22) found it necessary to import parental stocks from abroad. Local stocks had ceased to flower, evidently because of the curl. Thus, on its own, the earlier literature suggests a cyclic appearance of the curl. Varieties produced from seed might flourish during periods of decline in the curl, but succumb during the next outbreak. We now know, of course, that some varieties have survived many years in windswept coastal regions, where aphid vectors were virtually absent.

Potato Leaf Roll

Leaf Roll—A Sequence to the Curl—1905 to 1912

It is during the 20th century that we have obtained the strongest evidence for a longterm PLRV cycle. There have been three major outbreaks of the leaf roll virus in the northern seed-growing areas during the present century, and significantly, they have been virtually in phase-sequence with the earlier outbreaks of the curl (7).

The leaf roll disease, as we know it, was described first by Otto Appel at a meeting of German botanists in 1905. According to Pethybridge (36), the disease was widespread throughout most of northeastern Europe during the next several years. Appel thought that *Bladrollkrankheit* was one of the conditions included under the term *Krauselkrankheit*. Apparently, the curl had not been common for a time, as he remarked that it had been overlooked in recently preceding years. As references to curl, we note that he cited papers of 1854, 1875, and 1879.

Unravelling the Complex

For some years the leaf roll disease was confused with that caused by *Fusarium* spp. of fungi, even by Appel. But several workers noticed the absence of fungus in what Spieckerman (Refs. in 36) referred to as "typical leaf roll."

About this time, there began an international exchange of visits. W.A. Orton, from the United States, toured Europe in 1911 (34), and Appel visited the U.S. in 1914 (3). The two, in company with others, made a summer tour of the U.S. potato growing districts. Meanwhile, P.A. Murphy arrived in Canada and E.J. Wortley took up a post in Bermuda. As a result, North America was spared most of the confusion over any connection between *Fusarium* and leaf roll. In addition, Orton found in Germany, a disease-like condition that apparently had been overlooked up to this time. He termed it *mosaic*. On his return to America, he found mosaic to be widespread there, too. We note in his account of these diseases, the intuitive statement: ". . . the thought suggests itself that there have been periods, or cycles, of decline in potato varieties, followed by the rejuvenation due to the introduction of new sorts. It may be that such a period is now beginning, as manifested by the appearance of leaf roll and similar troubles during recent years."

Seed Potato Certification

Rather, the "decline" had virtually run its course, and the "rejuvenation" was about to take a new form. The method was not entirely new, however, for seed potato inspection and certification—with the attendant practices of roguing, the ascendancy of certain geographical areas, and winter testing of seed lots—borrowed heavily from the knowledge gained late in the 18th century. The standards for such programs in North America were established initially at a meeting in Philadelphia in December 1914 (39). It was attended by representatives from 12 of the states, as well as from the U.S. Department of Agriculture, Canada, Germany, and Ireland.

Ostensibly the leaf roll outbreak was brought under control in Germany after seed potato inspection was instituted in 1912 under the direction of Appel (4, 36). We should note that Putsch and Bertuch (Refs. in 6) wrote, almost 100 years earlier, in 1819: "The disease passes over into the tubers and even after careful cultivation (roguing), it will not disappear before the fourth or fifth generation." Presumably, they were describing events following 1812, as the disease could now be brought under control.

About 1912, there had been some trouble in America when Canadian seed shipped to Bermuda was found to contain alarming quantities of leaf roll. Canadian growers at the time were apparently unaware of the problem. But after some collaboration between Wortley and Murphy (32, 60, 61), the matter was cleared up under a new inspection scheme established by Murphy in Canada.

The Potato Leaf Roll Virus (PLRV)

It was left to Quanjer and his Dutch coworkers in 1916 to show that leaf roll was an infectious disease (37). At first, efforts were made to demonstrate soil transmission. But after this resulted in some inconsistencies, Oortwyjn-Botjes determined that the infectious principle, shortly to be termed a *virus*, was transmitted by aphids (33). Quanjer used the name *phloem-necrosis* for the disease, but common usage has remained with "leaf roll." It has been shown that PLRV required a relatively prolonged feeding period both for acquisition and transmission, and could be retained for days by the aphid vector. Thus transmission of the virus was virtually limited to potato-colonizing aphids, and *M. persicae*, has been considered the major vector (26).

PLRV Strikes Again—1937 to 1944

Whether it was due to the new inspection service, or simply the cyclic effect, the leaf roll disease went into decline in the northern seed-growing areas through the 1920s and early 1930s (7, 12, 32). The field, or rather fields, were virtually left to mosaic. After a prolonged struggle, the mosaic disease, too, was brought under reasonable control in the mid-1930s. But, just as this success was being savored by the inspectors, PLRV struck once more.

A survey of potato aphids had been undertaken in New Brunswick, Canada in 1934 (1). Four aphid species were common on potato plants in

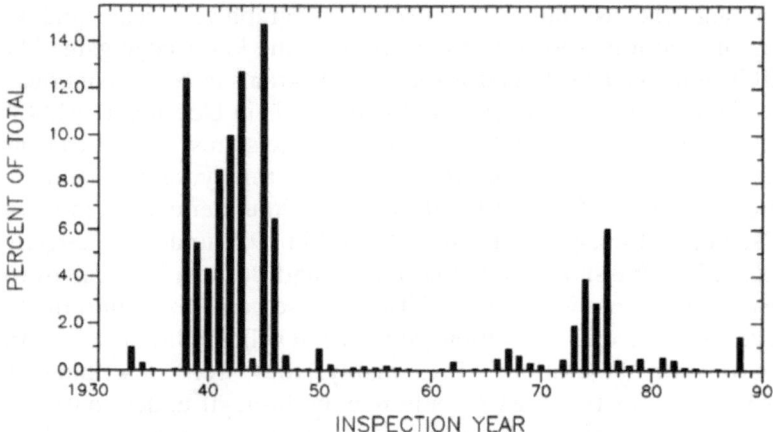

FIGURE 3.1. Relative area of seed potatoes refused certification due to excess leaf roll (PLRV), New Brunswick, 1933 to 1988. (Based on area passed plus area rejected for PLRV, combined; and converted to "percentage of the whole"—total of all bars adds to 100%. Area rejected for other reasons not included, as PLRV data is not available.) Note that this inspection data largely reflects spread during the previous year. (Updated from Ref. 7.)

the Saint John valley, but relatively scarce in the cooler Restigouche valley in the northeast of the province. These were similar to aphids found in Maine by Patch. The modern names are: *M. persicae*, the green peach aphid, *Macrosiphum euphorbiae*, (Thom.), the common potato aphid, *Aphis nasturtii*, Kalt., the buckthorn aphid, and *Aulacorthum solani* Kalt (Patch), the comparatively rare foxglove aphid. In 1936, it was evident that changes were taking place in the distribution of the species. All species were spreading northward, and the incidence of *M. persicae* was increasing, and so, as it developed, was the incidence of PLRV. It was not until midsummer, 1938, that the shocked inspectors realized what had happened. Over 30% of the New Brunswick seed crop was rejected for leaf roll in excess of the regulation 1% (Fig. 3.1).

There was a decline of PLRV during 1938 and 1939, but a renewed spread in 1940. Rejections for excess leaf roll increased in 4 of the next 5 years, reaching 35% of the area inspected in 1945. Just as matters appeared out of control, the epidemic quietly subsided during the next 2 years. With the advantage once more on their side, the inspection services and growers in New Brunswick quickly cleaned up the crop.

PLRV, *Myzus persicae*, and Climate

Events in the Aroostook County potato-growing area in northern Maine closely paralleled those just described in New Brunswick. Simpson and Shands (49) detailed the life-histories of the potato-colonizing aphids.

They noted that small numbers of *M. persicae* could overwinter as eggs on Canada plum (*Prunus nigra*).

But this aphid was less well adapted to the climate of the northern seed growing areas than was, for example, *A. nasturtii*. In this area, *M. persicae* was near the northern limits of its range, and appeared to be important only in recurring cycles. Simpson and Shands mentioned no period, but they noted that major populations of *M. persicae* and outbreaks of PLRV occurred in years with above normal temperatures (49). This was apparently so in most of the years between 1937 and 1942 (12) and evidently again in 1944 (see Table 1 in Ref. 18). Smith et al. (50) have confirmed that warm temperature, particularly early in August, is correlated with the spread of PLRV in northern Maine.

Meanwhile, *M. persicae* regularly occurred in southern Maine, where Schultz and Folsom (47) found the spread of PLRV to be higher than in the north. Spread of PLRV became progressively more severe further south on Long Island and in Virginia.

Gorham (23) found some *M. persicae* that apparently had survived in the egg stage on Canada plum in New Brunswick during the winter of 1940–41, at the time of the leaf roll epidemic. During surveys from 1948 to 1958, however, MacGillivray (30) could find only one such colony; and another was found in Maine near the New Brunswick border in 1971, a year of renewed PLRV spread. Evidently, overwintering of this aphid so far north was not a common occurrence in off-cycle years, although Canada plum occurs in numerous locations near potato fields in the seed growing area (24).

PLRV—Another Epidemic—1971 to 1975

In a virtual recurrence of similar events of the previous two centuries, a decline of 25 years, 1946 to 1970, ensued before the next major outbreak of PLRV (Fig. 3.1). On this occasion, we have records from both sides of the Atlantic (7). There was active spread of PLRV during 1971 to 1975 in the Maine–New Brunswick area and about 1971 to 1976 in Britain and northwestern Europe. The decline, again, was sudden.

The Range of *M. persicae*

The epidemic of PLRV in the 1970s once more raised alarm, and there was a renewed activity in the assessment of aphid populations. From the results of these and earlier studies both here and in Europe, I have ocncluded that the epidemiology of PLRV in northern seed–growing areas on both continents revolves about a periodic expansion–contraction or north–south shift effect in the seasonal population of *M. persicae* (7).

Although considerable individual efforts have been applied to studies of aphids in the Maine–New Brunswick area, there is genuine need for a continuing cooperative survey similar to Euraphid (53). Ideally, it should cover the entire Atlantic seaboard from northern New Brunswick to Florida. For *M. persicae*, this would range from the aphids' northern limits to an area where its survival is entirely anholocyclic.

On a less ambitious scale, we should at least have trapping and field surveys in the northeast region with contemporary circulation of the results. The survey area should extend from the Monmouth area in southern Maine—the site of Schultz and Folsoms' trials (47)—to northern Aroostook County, with a similar effort in the area along the Saint John River in New Brunswick.

During the early 1980s, when PLRV was relatively scarce in New Brunswick potatoes, small amounts of PLRV regularly could be found in the area near Centreville—opposite Bridgewater, Maine. Relief maps of Maine and New Brunswick show Bridgewater and Centreville on the northern edge of a basin drained by the Kennebec, Penobscott, and upper St. Croix rivers in Maine and the Maduxnekeag—a tributary of the Saint John—in New Brunswick. The scattered data that we have of *M. persicae* and PLRV suggest that there is a gradient in the incidence of both from south to north.

Moreover, there is a range of high hills, anchored by Mt. Katahdin in Maine and Mt. Carleton in New Brunswick that appear at least partly to shield the northermost areas of the potato belt from migration of southern aphids. Even during epidemic times, there is considerably less PLRV in the northern part of Aroostook County in Maine, and its counterpart, Victoria County in New Brunswick, than in areas just to the south.

PLRV Epidemics—A 32-Year Cycle

With actual data from the two epidemics, 1937 to 1945 and 1971 to 1976, as well as some years of decline that followed each; and partial records from 5 of the earlier epidemics, the author sought factors in common (7).

Although the epidemics might appear to break out suddenly in one season or build up over several, a repetitive factor has been a final flare-up or climax during a particularly warm season. This was typically followed by an abrupt decline after one or more severe winters (7, 54). Such consecutive winters as 1946–47 and 1947–48, 1975–76, and 1976–77 drastically reduced the populations of *M. persicae* in the respective following summers (see Tables 1 and 2 in Ref. 7).

A 32-year cycle was evident, with the climax years approximately **1784, 1812, 1844** to **1848, 1878, 1912, 1944**, and **1975** to **1976**. In a search for any possible climatic cycle effect, I noticed an odd near-coincidence between these climax years and the reputed minima of periodic droughts in the western Sahel of the SubSaharan region of North Africa (16, 17). It appears that this is not a chance occurrence, for Vines (59) has detected three distinct, but concurrent rainfall cycles in different parts of the globe, including northeast United States and eastern Canada. These cycles of 6 to 7, 11-, and 16-years tend to coincide (at multiples of 5×, 3×, and 2×, respectively) and reinforce each other at 30-plus year intervals. Vines thought that this reinforcing effect might account for the Sahel droughts, which he described as one of the major climatic anomalies of the past century. The 11-year drought cycle is in phase with sunspot minima, thus our climax years have been close to the minima of each third sunspot cycle since 1812. The climax of 1784 was also at a minimum, but two relatively long cycles earlier (7).

The subject of sunspots and other extraterrestrial influence on weather and climate were discussed briefly in a previous paper (7). It is sufficient here, to say that after a decline in acceptance for several decades, these have once more become respectable subjects for scientific research in the 1980s.

PLRV in a Warming Climate

As a final comment (Refs. in 7) I had noted reports of sharply increasing global temperature in recent years (29), and suggested that if this were to continue, PLRV epidemics might become more frequent or prolonged. Evidently temperatures have increased since the 1970s, because the warmest years of this century have occurred in the 1980s. And locally, populations of *M. persicae* have been high during 1985 to 1988. Another outbreak of PLRV—at least one of moderate severity—has resulted. The season of 1988 in New Brunswick had several characteristics of a climax year, that is, warm and dry, with large populations of *M. persicae*. Observations in 1989 indicate that there has once more been a sharp decline in spread of PLRV. Although the winter of 1988–89 was not particularly cold overall, there were several abrupt shifts between mild and severe temperatures.

I mentioned, too (Refs. in 7), that there had been a minor outbreak of PLRV in 1965 and 1966, as a prelude to that of 1971 to 1975. Sunspot minima for these times were 1964.7, 1975.7, and 1986.6 (year.month) (55). Thus the climax years of each of these outbreaks, 1965 to 1966, 1971 to 1975, and 1985 to 1988 were close to a sunspot minimum. If the climate is indeed warming, the question arises: Is the 11-year drought cycle, without need for amplification by Vines' 16-year cycle (59), now capable of triggering a PLRV epidemic?

Potato Mosaic

Mosaic and the Curl

When Schultz and his coworkers (48) demonstrated in 1919 that mosaic was an infectious disease, and that the principle, as with that of leaf roll, could be transmitted by aphids, the possibility of double infections involving a mosaic virus and PLRV became a reality.

We know that "mild mosaic," now ascribed to PVA, figured at least partly in Orton's (34) first description of mosaic, because he found the Irish Cobbler variety free from the disease in parts of northern Maine. This variety is highly resistant to PVA. Mild mosaic was still widespread in Maine in 1919 when the U.S. Department of Agriculture set up a breeding program at Presque Isle (52); and this was probably so of New Brunswick, as well.

Orton (34), Murphy (32), and Schutz and Folsom (46, 47) added to the picture by describing several other forms of mosaic, including rugose mosaic, crinkle, and streak. Atanasoff (5) described another variation, stipple-streak. Basically, these were a miscellany of current-season or secondary symptoms, frequently differing with the variety or age of the potato plant involved, of what we now know to be a number of different strains of PVY. If experience during the 20th century is an indication, the mosaic disease was probably present in one form or another throughout earlier times. Combined with periodic outbreaks of PLRV, this could account for the cyclic behavior of severe forms of the curl. This double infection is probably not common with crops planted with todays' "elite" seed stocks, but the effect on certain cultivars can be extremely severe. The term *curl* could aptly apply.

The 9-Plus Year Mosaic Cycle in New Brunswick

Potato inspection records in New Brunswick going back to 1919 (Fig. 3.2), show a waxing and waning of the mosaic disease. The incidence was high at this time, and the initial struggle to control it lasted through the 1920s. But the mosaic epidemic faltered at the end of the decade. The incidence peaked briefly once more, as reflected by an increase in field rejections for excess mosaic in 1932 and 1933. There was then a sharp decline in the mid-1930s.

The peak of 1932 began a series of ups and downs in the rate of rejections by field inspectors for excess mosaic (Fig. 3.2). The individual peaks were highlighted by 1 or 2 years standing above other contiguous years. Where there were two nearly equal, I select the first. We should note that much of the mosaic seen during the inspection year reflects spread during the previous year—which I term the *epidemic* year. If we take the epidemic years during which the spread of the mosaic viruses or the "infection pressure" was highest, then we have the sequence: **1931, 1941, 1949, 1959, 1968, 1978,** and **1987**. Though there are gaps in our early data, the years 1920 to 1928 appear to form a virtual plateau, with sufficient spread to produce 25 to 33% rejections each year—with the significant exception of 1921, when epidemic spread of the mosaic viruses resulted in

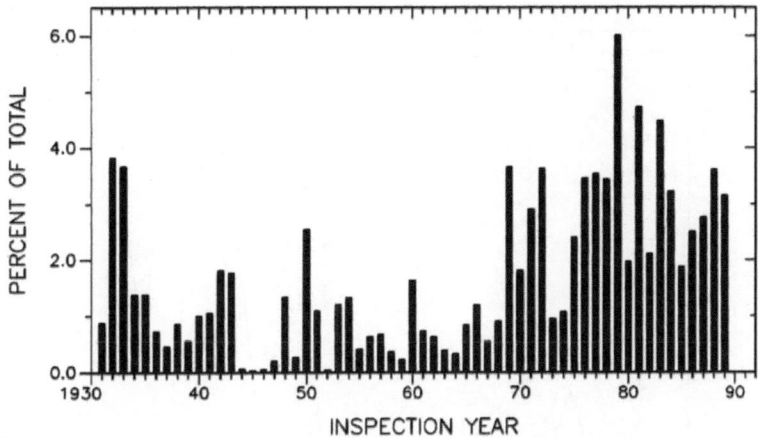

FIGURE 3.2. Relative area of seed potatoes refused certification due to excess mosaic, largely due to potato virus Y (PVY), New Brunswick, 1931 to 1989. (Based on area of PVY-susceptible cultivars passed plus an area rejected for mosaic, combined; and converted to "percentage of the whole"—total of all bars adds to 100%. PVY-resistant cultivars, categories 4 and 5 of Ref. 9 excluded, as there have been virtually no rejections for mosaic amongst these. Areas rejected for other reasons have also been excluded due to lack of data re: mosaic.)

55% rejections in the following year. With an average interval between peak years of 9.3 years, 1931 to 1987 (9.4, if we take 1921 to 1987), this is clearly out of phase with either an 11- or 32-year PLRV cycle.

The Mosaic Cycle and the Canada Wildlife Cycle

However, another well-known cycle comes to mind. This is the Canada wildlife cycle—a 9-plus-year cycle reputed to affect a number of animals, birds, fish, and insects in the Boreal forest region of the northern hemisphere (2, 13, 19). This cycle is well documented for the past 160 years as the Canada lynx cycle (13, 18, 31). There is a plausible connection between the the lynx and PVY cycles. The Canada lynx preys largely on the snowshoe hare, which depends on foliage of small trees and shrubs for food and shelter. (Thus inreality there is a Canada hare-lynx cycle.) Many of the same trees and shrubs also serve as overwintering hosts for eggs of different aphid species. And many of these aphids, although not colonizing potato plants, visit potato fields, and probe leaves in search of suitable hosts—a behavior ideally suited to transmission of the nonpersistent PVY (25). Fox (19) suggests that the lynx cycle is driven by the new tree growth that follows drought and forest fires. He shows that major fires in Canada tend to occur in a 9 to 10 year cycle.

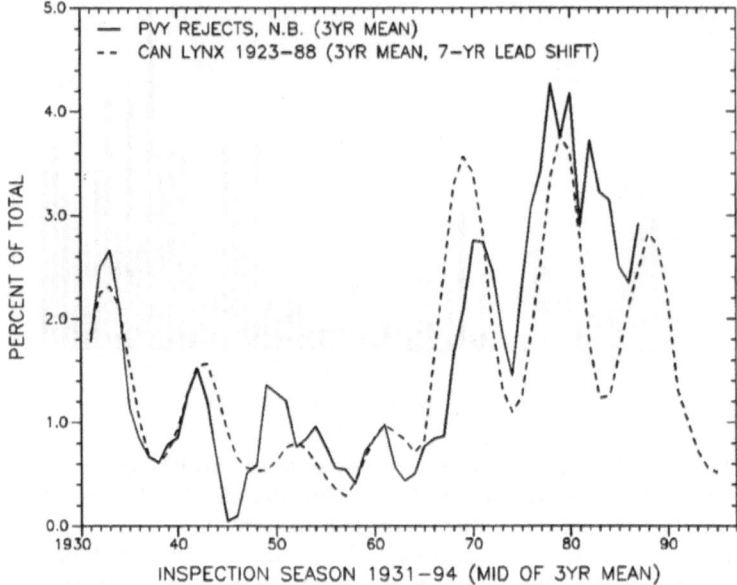

FIGURE 3.3. Comparison of peaks of the New Brunswick mosaic or PVY cycle, 1931 to 1988 (Fig. 3.2), with those of the Canada lynx cycle 1924 to 1988. (Both sets are 3-year running means converted to "percentage of the whole," to bring amplitudes into a common range. Canada lynx data moved 7 years to the right to represent a proposed "lead time."

The mosaic or PVY cycle is compared with the Canada lynx cycle in Figure 3.3, where I use 3-year running means. To bring the amplitudes into a common range, I convert items in the y-scales into "percentage of the whole", and additionally, I move the Canada lynx cycle 7 years to the right. We note a downward trend in both cycles from the 1920s, with a reversal of the lynx series in the late 1950s and of the PVY cycle in 1965. We cannot be certain, at this stage, that the lynx and PVY cycles are similarly driven. But if they are, it is fair to speculate that aphids thrive on older and larger trees than do hares. In any case, these moves produce a good match of 9-plus-year peaks and an excellent one in the long-term trends.

There is an intriguing prospect that we might use the lynx cycle, with its 7-year lead—6 years if we focus on the epidemic years of the PVY cycle, rather than the inspection years—as a guide to the future trend in the PVY cycle. The lynx harvest after a high for several cycles in 1972, was lowest since 1950 in the years 1986 to 1988 (51). Meanwhile, the trend in the PVY cycle after a long-term epidemic year peak in 1978, also appears to be down.

The British Sugarbeet Yellows Cycle

Another cycle that has some characteristics in common with the PVY cycle is the sugarbeet yellows disease cycle in Britain. Here, too, there are individual high years at approximately 9-year intervals. There were highs in 1949, 1957, and 1974, plus two weak or missing highs about 1966 and 1983 (21, 28, 41). Gibbs initially suggested that the peaks were in phase with high sunspot activity and thus to be expected at 11-year intervals. At the time of the major peak in 1974, however, the sugarbeet yellows cycle had moved out of phase with sunspot maxima, and was close to a sunspot

The Biennial Rhythm in PVY Epidemics

The PVY cycle in New Brunswick has actually proved to be more complex (8).

The spread of PVY into initially virus-free plants of the cultivar Russet Burbank was monitored in a field trial at Fredericton, New Brunswick. Infected plants were counted at intervals. The successive counts during each season were used to produce disease progression curves. But, based on disease appearance, rather than the time of infection, these curves were subject to an incubation-period time-lag of two or more weeks.

In a parallel trial, 1978 to 1988, small plots were exposed in sequence, two plots at a time, for intervals of 7 to 14 days, while being covered with saran-screen cages at other times during the season. Tubers were indexed in the greenhouse over the winter to identify those that had become infected. These results were used to produce current infection curves, based only on those infections that eventually reached the tubers.

The results of these two trials were in agreement in that there were two distinct types of season: (a) "high," typically with early initial infections and with high-cumulative infection with PVY; and (b) "low," with late initial infections and low-cumulative infection.

A biennial rhythm was evident from 1974 to 1982, with even years high and odd years low. A transition then took place, for 1983 and 1984 were both low, and the rhythm continued with the odd years, 1985 and 1987, high and the even years, 1986 and 1988, low. Results of the New Brunswick Florida test for the susceptible cultivar Russet Burbank follow much of the same pattern—and, significantly, show another double-low in 1972 and 1973 (8).

This biennial rhythm appears to have had an important bearing on the health of the New Brunswick seed potato crop. There has been a significant spread of PVY in only seven of the past 17 years, with no two in succession. Hille Ris Lambers (25, 26) reported a somewhat similar biennial rhythm in aphid populations in the Netherlands. Populations were high in uneven

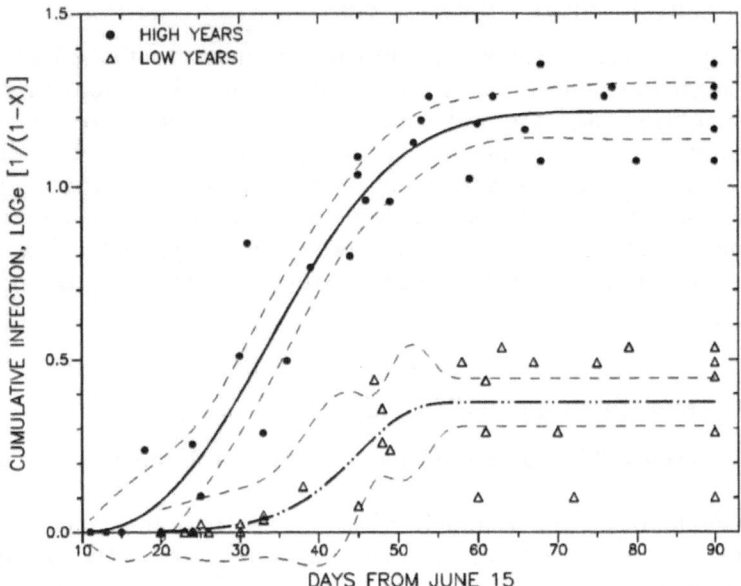

FIGURE 3.4. Cumulative infection with potato virus Y in caged plots of cultivar Russet Burbank, successively exposed and re-covered in field trials, Fredericton, New Brunswick. Data converted from percentages: $y = \text{Log}_e [1/(1 - x)]$, after Vanderplank (56). Solid curve, Mean for "High spread" years, 1978, 1980, 1982, 1985, and 1987; Dash-dot curve, Mean for "Low spread years," 1979, 1981, 1983, 1984, and 1986, Broken lines, confidence belts for means.

years during the 18-year period of 1936 to 1954. He saw in this, the workings of a predator–prey cycle.

Seasonal Spread of PVY and Mature Plant Resistance

My own open-field monitor trials, mentioned above, gave an excellent measure of cumulative infection for the season. But the cage-exposure trials gave probably the best picture of within-season trends. Infections that eventually reached the tubers accumulated in the form of a sigmoid curve (Fig. 3.4). Most such infections in the high years occurred between July 7 and August 1, whereas in low years—at a lesser rate—between July 20 and August 1. A decline in the infection rate, due to mature plant resistance appeared to begin late in July. Few infection sites established after August 1 were likely to result in the virus penetrating to the tubers. This important factor normally does not show in trials where only the field plants are monitored, as in my first trial.

There is evidence, however, that in some years at least, aphids continue to acquire PVY and transmit it as late as mid-September. Young, potted potato plants set out September 1 to 15 have become infected to the extent of 25%. Thus tests conducted with immature bait plants may give a misleading picture of an epidemic.

The Persistent PLRV and Nonpersistent PVY—Different Cycles, Different Aphid Vectors

My own observation is that the difference between the PLRV and PVY cycles stems from differences in their modes of transmission.

Transmission of the persistent-type PLRV depends almost exclusively on the potato-colonizing aphid, *M. persicae*, which overwinters poorly in the north. Its appearance is subject to the climatic influences described above. *M. persicae* arrives in northern potato fields relatively late in the season. Unless the aphids have come directly from potato fields further south, there will likely be need of additional time for a new generation of PLRV-infective aphids to develop on potato plants before there can be significant spread of this virus. *And time is of the essence*, for a week or two difference can be extremely important relative to mature plant resistance (10).

Potatoes are planted in the New Brunswick–Maine "potato belt" about May 20 to 31, and emerge June 10 to 20. Mature plant resistance to PLRV in the major cultivar Russet Burbank set in shortly after mid-August. For PVY, such resistance appears earlier, as noted above.

Some of the "native" northern aphids capable of transmitting PVY make their appearance in late June or early July, when potato plants are most susceptible. At the earliest stage, spread of PVY is probably limited by the small size of the potato plants, as these aphids do not colonize potatoes. Nevertheless, I have detected spread of PVY in "high" years before July 1. It is these native aphids that appear to be most subject to the 9-plus-year cycle and to the biennial rhythm.

Hille Ris Lambers (27) considered aphids such as *A. nasturtii, Aphis fabae Scop.*, and *Rhaphalosiphum padi* L. to be proven, efficient vectors of PVY. He felt that many other aphids should be held suspect as vectors. Negative results in laboratory trials need not mean that the aphid could not transmit in the field. He had found this to be so with *R. padi*, which could transmit PVY under simulated field conditions.

M. persicae is an efficient vector of PVY, and undoubtedly has a role in the spread of this virus in New Brunswick. But it does not appear to reach its full potential under our conditions. From 1974 to 1983, the first trapping of *M. persicae* in New Brunswick averaged July 24 (11). And the earliest trappings were July 18, 19, and 22, during the years 1981, 1975, and 1977, respectively—all of them "low" years in my trials. With this relatively late arrival, *M. persicae* normally has only a short period for effective action.

This is so, despite the fact that the first arrivals can commence to spread PVY immediately. We know, for example, that 1976 was a high year for the spread of PVY in New Brunswick, but *M. persicae* was virtually absent, and spread of PLRV was extremely low. (1). The years 1978, 1980, and 1982 were also years of high-PVY spread, but populations of *M. persicae* and spread of PLRV were quite modest. Conversely, when populations of *M. persicae* and spread of PLRV were high during 1986 and 1988, the spread of PVY was relatively both late and low.

Ryden et al. (42) in Sweden reported experiences similar to the above. Gabriel et al. (20) in Poland, and Van Harten (57) and Van Hoof (58) in the Netherlands, also found reason to consider species other than *M. persicae* important vectors of PVY. Schultz and Folsom, as early as 1923, found that the buckthorn aphid (*Aphis abbreviata*—now termed *A. nasturtii*) could transmit the mosaic viruses (PVA and PVY). Harrington et al., (25) in Britain, trapped live aphids near potatoes heavily infected with PVY. When placed directly on healthy tobacco host plants, 147 aphids of 20 different species transmitted naturally acquired PVY to the tobacco. *M. persicae* accounted for 11 (7.5%) of these infections.

Acknowledgments. I have received invaluable help with this chapter from my colleague, Merv C. Clark, and the station librarian, Richard Anderson. I am also indebted to E.M. Smith for the New Brunswick seed inspection records.

References

1. Adams, J.B., 1963, Potato aphid studies in New Brunswick, 1924–1962, *Aphidologists Newsletter*, St.Jean, Quebec **2** (2):1–3.
2. Anderson, R.M., and May, R.M., 1981, The population dynamics of microparasites and their invertebrate hosts, *Phil. Trans. R. Lond. Soc. B.* **291**:454–524.
3. Appel, O., 1915, Leaf roll diseases of the potato, *Phytopathology* **5**:139–148.
4. Appel, O., 1934, Vitality and vitality determination in potatoes, *Phytopathology* **24**:482–494.
5. Atanasoff, D., 1922, A study into the literature on stipple-streak and related diseases of potato, *Meded. Landb. Hoogesch, Wageningen* **26**:1–52.
6. Atanasoff, D., 1922, Stipple-streak disease of potato, *Meded. Lanbouhoogesch., Wageningen.* **24**:1–32.
7. Bagnall, R.H., 1988, Epidemics of potato leaf roll in North America and Europe linked to drought and sunspot cycles, *Can. J. Plant Pathol.* **10**:193–202.
8. Bagnall, R.H., 1989, The potato mosaic, virus Y, and latent virus S cycle, *Agric. Canada, Res. Branch, Ann. Rept.* **1988**:177–178.
9. Bagnall, R.H., and Tai, G.C.C. 1986, Field resistance to potato virus Y in potato assessed by cluster analysis, *Plant Dis.* **70**:301–304.

10. Beemster, A.B.R., 1987, Virus translocation and mature plant resistance in potato plants, DeBokx, J.A., and Van der Want, J.P.H. (eds): in Viruses of Potatoes and Seed Potato Production, 2nd. ed. PUDOC, Wageningen, pp. 116–125.

11. Boiteau, G., and Parry, R.H., 1985, Monitoring of flights of green peach aphids, *Myzus persicae* (Sulzer), in New Brunswick potato fields by Yellow pans from 1974 to 1983: Results and degree-day simulation, *Amer. Potato J.* **62**:489–496.

12. Bonde, R., Schultz, E.S., and Raleigh, W.P., 1943, Rate of spread and effect on yield of potato virus diseases, *Maine Agric. Exp. Stn. Bull.* **421**:1–28.

13. Bulmer, M.G., 1974, A statistical analysis of the 10-year cycle in Canada, *J. Animal Ecol.* **43**:701–718.

14. Davidson, W.D., 1928, A review of literature dealing with the degeneration of varieties of the potato, *Econ. Proc. Roy. Dublin Soc.* **2**:331–389.

15. Davidson, W.D., 1934, History of potato varieties, *J. Dept. Agric. Repub. Ireland* **33**:57–81.

16. Faure, H., and Gac, J.-Y. 1981, Will the Sahelian drought end in 1985? *Nature* **291**:475–478.

17. Faure, H., and Gac, J.-Y. 1981, Faure and Gac reply (to Palutikoff, Gough, and Farmer), *Nature* **293**:414.

18. Folsom, D., and Stevenson, R.V., 1946, Resistance of potato seedling varieties to the natural spread of leaf roll, *Amer. Potato. J.* **23**:247–264.

19. Fox, J.F., 1978, Forest fires and the snowshoe hare-Canada lynx cycle, *Oecologia (Berlin)* **31**:349–374.

20. Gabriel, W., Kostiw, M., and Wislocka, M., 1975, Comparaison de plusieurs méthodes d'estimation de la quantité de pucerons vecteurs de virus, pour la prévision d'infection par virus des tubercules de pomme de terre, *Potato Res* **18**:3–15.

21. Gibbs, A.J., 1966, A possible correlation between sugar beet yellows incidence and sunspot activity, *Plant Pathol.* **15**:150–152.

22. Goodrich, C.E. 1884, The potato. Its diseases—with incidental remarks on its soils and culture, *Trans. N.Y. State Agric. Soc.* 1883, **23**:103–134; with memoir by J.P. Gray, pp. 135–139.

23. Gorham, R.P., 1942, The progress of the potato aphid survey in New Brunswick and adjacent provinces, *Ent. Soc. Ontario, 72nd Ann. Rept.* **1941**:18–20.

24. Gorham, R.P., 1943, The distribution of *Prunus nigra* Ait. in New Brunswick, *Acadian Naturalist* **1**:43–46.

25. Harrington, R., Katis, N., and Gibson, R.W. 1986, Field assessment of the relative importance of different aphid species in the transmission of potato virus Y, *Potato Res.* **29**:67–76.

26. Hille Ris Lambers, D., 1955, Potato aphids and virus diseases in the Netherlands, *Ann. Appl. Biol.* **42**:355–360.

27. Hille Ris Lambers, D., 1972, Aphids: their life cycles and their role as virus vectors, DeBokx, J.A., (ed): in Viruses of potatoes and seed potato production, PUDOC, Wageningen, pp. 36–55.

28. Heathcote, G.D., 1986, Virus yellows of sugar beet, Mclean, G.D., Garrett, R.G., and Ruesink, W.G., (eds): in Plant Virus Epidemics. Monitoring, Modelling, and Predicting Outbreaks, Academic Press, Sydney pp. 399–417.

29. Jones, P.D., Wigley, T.M.L., and Wright, P.B. 1986, Global temperature variations between 1861 and 1984, *Nature* **322**:430–434.
30. MacGillivray, M.E., 1972, The sexuality of *Myzus persicae* (Sulzer), the green peach aphid, in New Brunswick. *Can. J. Zool.* **50**:469–471.
31. McKee, R.K., 1964, Virus infection in South American Potatoes, *Eur. Potato J.* **7**:145–151.
32. Murphy, P.A., 1921 Investigation of potato diseases, Canada *Dept. Agric. Exp. Farms Bull.* 44 (2nd ser.):1–86.
33. Oortwijn-Botjes, J.G., 1920, The leaf roll disease of the potato plant, [In Dutch], Dissertation, Agric. Univ. Wageningen, 136 pp.
34. Orton, W.A., 1914, Potato wilt, leaf-roll, and related diseases, *U.S.D.A. Bull.* **64**:1–48.
35. Pethybridge, G.H., 1924, Potato leaf roll, *J. Ministry Agric. Gr. Brit.* **31**:863–869.
36. Pethybridge, G.H., 1939, History and connotation of the term "Blattrollkrankheit" (leaf-roll-disease) as applied to certain potato diseases, *Phytopathol. Z.* **12**:283–291.
37. Quanjer, H.M., van der Lek, H.A.A., and Oortwyjn-Botjes, J., 1916, Nature, mode of dissemination and control of phloem necrosis (leaf roll) and related diseases, [In Dutch], *Meded. Rijks. hoog. land-, tuin-boschbouwschool* **10**:1–90.
38. Quanjer, H.M., 1922, New work on leaf-curl and allied diseases in Holland, *Rept. Internat. Potato Conf., Roy. Hort. Soc.,* London, Nov. 16–18, **1921**:127–152.
39. Rieman G.H. Ed., 1956, Early history of potato seed certification in North America, 1913–1922, Potato Handbook *Potato Assn. Amer.,* New Brunswick, N.J., 1956:6–10.
40. Rodriguez, A. and Jones, R.A.C., 1978, Enanismo Amarillo disease of *Solanum andigena* potatoes is caused by potato leaf roll virus, *Phytopathology* **68**:39–43.
41. Rothamstead Experimental Station, 1984–1987, *Ann. Repts., Agron. and Crop Physiol. Div.*
42. Ryden, K., Brishammar, S., and Sigvald, R., 1983, The infection pressure of potato virus Y° and the occurrence of winged aphids in potato fields in Sweden, *Potato Res.* **26**:229–235.
43. Salaman, R.N., 1922, Degeneration of potatoes, *Rept. Internat. Potato Conf.,* London: Roy. Hort. Soc., Nov. 16–18, 1921:79–91.
44. Salaman, R.N., 1926, Potato Varieties, University Press, Cambridge, 378.
45. Salaman, R.N., 1948, Some notes on the history of curl. *Tijdsch. Plantenziek.* **55**:118–128.
46. Schultz, E.S. and Folsom, D., 1920, Transmission of the mosaic disease of Irish potatoes, *J. Agr. Res.* **19**:315–337.
47. Schultz, E.S. and Folsom, D., 1925, Infection and dissemination experiments with degeneration diseases of potatoes, Observations in 1923, *J. Agr. Res.* **30**:493–528.
48. Schultz, E.S., Folsom, D., Hildebrandt, F.M., and Hawkins, L.A. 1919, Investigations of the mosaic disease of the Irish potato, *J. Agr. Res.* **17**:247–273.
49. Simpson, G. and Shands, W.A., 1949, Progress on some important insect and

disease problems of Irish potato production in Maine, *Maine Agr. Expt. Stn. Bull.* **470**:1–50.

50. Smith, O.P., Storch, R.H., Hepler, P.R., and Manzer F.E., 1984, Prediction of potato leaf roll virus disease in Maine from thermal unit accumulation and an estimate of primary inoculum, *Plant Dis.* **68**:863–865.
51. Statistics Canada, 1919–1988, Fur production, Cat 23–207, Annual.
52. Stevenson, F.J. and Akeley, R.V., 1953 Control of potato diseases by disease resistance, *Phytopathology* **43**:245–253.
53. Taylor, L.R., 1983, Euraphid: synoptic monitoring of migrant vector aphids, Plumb, R.T. and Thresh, J.M., (eds): in Plant Virus Epidemiology, Blackwell, Oxford, pp. 133–146.
54. Turl, L.A.D., 1983, The effect of winter weather on the survival of aphid populations on weeds in Scotland, *EPPO Bul.* **13**:139–143.
55. U.S. Department of Commerce, 1989, Monthly sunspot numbers Jan. 1947–Apr. 1989, *Solar-Geophys. Data.* Boulder, Col., Apr. 1989: p. 14.
56. Vanderplank, J.E., 1972, Plant Disease: Epidemics and Control, Academic Press, New York, 349 p.
57. Van Harten, A., 1983, The relation between aphid flights and the spread of potato virus Y^n (PVY^n) in the Netherlands, *Potato Res.* **26**:1–215.
58. Van Hoof, H.A., 1977, Determination of infection pressure of potato virus Y^N. *Neth. J. Plant Path.* **83**:123–127.
59. Vines, R.G., 1984, Rainfall patterns in the eastern United States, *Climatic Change* **6**:79–98.
60. Wortley, E.J., 1918, Potato leaf-roll: Its diagnosis and cause. *Phytopathology* **8**:507–529.
61. Wortley, E.J. and Murphy, P.A., 1920, Relation of climate to the development and control of leaf roll of potato, *Phytopathology* **10**:407–411.

4
Interactions Between Barley Yellow Dwarf Virus Infection and Winter-Stress Tolerance in Cereals

Christopher J. Andrews and Ramesh C. Sinha

Introduction

The productivity of a crop can be adversely affected by both abiotic and biotic factors. Some pathogens can kill infected plants and affect a crop so badly that it is uneconomic to harvest. Similarly, severe winter conditions or drought may ruin agriculturally important winter or summer crops affecting the economy of a country and the availability of food for both humans and livestock. Researchers in various parts of the world, therefore, have been involved in breeding crop cultivars that can resist damage due to harsh environments and various disease-causing agents. To achieve this aim, it is essential to understand the mechanism by which the crops are damaged by biotic and abiotic factors. At our Research Centre, pathologists have been investigating the consequences of interaction of viruses with cereals, while physiologists have focused their efforts examining the effects of winter conditions on the survival of cereals. However, very little work has been reported on the effects of virus infections on winter-stress tolerances in cereals. A few years ago, such studies were initiated by the late Dr. Y.C. Paliwal and the senior author at our Research Centre. Almost all the research on the effects of virus infections on winter-stress tolerances has been done using barley yellow dwarf virus (BYDV) and cereals, but some work has been reported also involving wheat spindle-streak mosaic virus (WSSMV). This chapter reviews the progress made on this subject. A brief description of properties, transmission, and epidemiology of BYDV and WSSMV has been included also in order to facilitate understanding of the effects of virus infections on winter-stress tolerances in cereals.

Christopher J. Andrews, Plant Research Centre, Agriculture Canada, Ottawa, Ontario, Canada, K1A 0C6.
Ramesh C. Sinha, Plant Research Centre, Agriculture Canada, Ottawa, Ontario, Canada, K1A 0C6.

Virus and Vectors

BYDV, a type member of luteovirus group (89), with small icosahedral virions about 25 nm in diameter contains a single-stranded RNA genome (mol. wt. 2×10^6 daltons). BYDV can be transmitted only by means of aphid vectors, presumably because it is host-tissue specific and multiplies primarily in phloem cells (48). The phloem cells of infected plants become necrotic, thereby disrupting the translocation of plant metabolites (49). Five variants of BYDV have been identified based on the specificity of their transmission by different species of aphid vectors and their comparative virulence on oats (35, 38, 80, 85). The isolate RPV is transmitted specifically by *Rhopalosiphum padi* (L.), RMV by *R. maidis* (Fitch), MAV by *Sitobion avenae* (Fab.) (= *Macrosiphum avenae*), SGV by *Schizaphis graminum* (Rondani), and PAV transmitted nonspecifically by *R. padi, S. avenae*, and *S. graminum*. Double-stranded RNAs (ds RNAs) have been isolated and identified from the plants infected with each of the five characterized isolates of BYDV (34). Based on the number and differences in the electrophoretic mobility of the isolated ds RNAs, BYDV isolates were divided into two distinct groups. Group 1 consisted of isolates MAV, PAV, and SGV and group 2 RPV and RMV. Five ds RNAs (mol. wt. of 3.6, 2.0, 1.2, 0.55, and 0.5×10^6 daltons) were observed from plants infected with group 1 isolates and only four ds RNAs (mol. wt. of 3.8, 1.6, 1.2, and 0.55×10^6 daltons) from plants infected with group 2 isolates. These differences agree with earlier evidence, based on serology (88) and cytopathology (39), and suggest that each of the BYDV isolates belongs to one of two distinct groups.

BYDV is transmitted by aphid vectors in a persistent manner. Once the aphid has acquired the virus, it can transmit the virus for several days or weeks (80). The circulative nature of the virus in its aphid vectors has been demonstrated (59). The virus has been recovered, using infectivity bioassays, from the gut, hemolymph, and salivary glands of aphids that had been fed on infected plants (68). BYDV does not appear to multiply in aphid vectors because it cannot be maintained in the aphids by serial passage of the virus from insect to insect. Also, the ability of aphids to transmit the virus declined after acquisition or injection of the virus into aphids, and the period of retention of inoculativity and transmission efficiency of aphids was dependent on the dose of virus injected or ingested (68). The ability of a plant virus to be transmitted by aphids is determined by the genetic constitution of the virus (43). The genetic traits of the vector also play a role in determining virus transmission but they are not well understood. Enough indirect evidence has accumulated to suggest that cell membrane receptor-mediated endocytosis may be involved in BYDV acquisition by aphid vectors. In an exhaustive review by Gildow (33), a hypothetical model was proposed to explain the mechanism of transmission of leutoviruses by aphid vectors. It was suggested that successful virus transmission requires virus recognition and binding to specific receptors that regulate virus uptake into the accessory salivary gland of aphids. The role of the accessory gland as a specific site for selective regulation of transmission of circulative viruses was first suggested by Harris and coworkers (41, 42), based on ultrastructural studies on pea enation mosaic virus in its aphid vector *Acyrthosiphon pisum*.

Mixed infections of plants by viruses in nature can play a key role in disease spread. Sometimes certain aphid species are able to transmit a virus from a plant only if it is also infected by a second virus. Such dependent transmission, discussed in detail by Rochow (82, 83), has been demonstrated for the MAV isolate of BYDV by *R. padi*, which does not transmit this isolate normally. However, in the presence of helper isolate RPV, *R. padi* could transmit MAV as well as RPV isolates from mixed infections (81, 86). It was suggested that in plants infected with both the isolates, the protein of RPV encapsidates the RNA of MAV isolate, which then can be transmitted by *R. padi* (81, 84, 90).

WSSMV particles are filamentous, measuring from 19 to 1975 nm in length and about 12.8 nm in diameter in leaf-dip preparations (94). In sections of infected wheat, however, the virus particles measured 18 to 20 nm in diameter with lengths exceeding 3000 nm (44, 56). The virus is sap-transmissible, and in soil it is transmitted by the fungus *Polymyxa graminis* (14). Biochemical properties, vector–virus relationships, and composition of WSSMV are not well known. Wheat (*Triticum* spp.) is the only known host of the virus. WSSMV affects winter wheat (*T. aestivum*) in Canada and the United States and durum wheat (*T. durum*) in southern France (93).

Epidemiology

BYDV is the most widespread cereal virus and causes a serious disease in oats, barley, wheat, and grasses in the temperate and subtropical world. Epidemiology of the virus has been well illustrated in a review by Plumb (72) entitled "BYDV—a global problem." In Canada, BYDV has been shown to be widely spread and has reached epidemic proportions several times. In 1969, the virus caused an estimated loss of 1.4 million bushels of barley in Manitoba (36). Another outbreak of the virus occurred in the same province in 1978 and the total estimated losses on bread wheat affected by the epidemic were 5.8 million bushels or 7% of the potential yield (37). In 1976, the BYDV epidemic was one of the most severe in eastern Canada (21), with yield losses of cereals ranging from 50 to 70% reported in many areas. An epidemic of the virus in winter wheat and barley crops in most parts of Ontario and Quebec in 1982 to 1983 infected about 50% of the winter wheat (67). The virus can be diagnosed by the distinctive symptoms it produces on indicator hosts, by vector transmission, and by serological means including highly sensitive tests such as enzyme-linked immunosorbent assay (ELISA) and immunosorbent electron microscopy (ISEM) (57, 61, 62, 64, 74, 87).

Grasses have been considered to be one of the main reservoir hosts for several BYDV variants. In the U.K., it has been suggested that at least one aphid vector species, *R. padi*, that develops mainly on grasses can carry the virus to the newly emerged autumn-sown cereals (73). In some areas of England, maize is the main source of BYDV carried by *R. padi* migrating during September to November (22). Plumb (72) stated that one of the most dangerous cropping practices is to follow a grass crop too quickly by a cereal. In such cases, viruliferous aphids can move from the dying infected grass to the emerging cereal crop that would become infected with BYDV then. Paliwal (63, 65) studied the role of perennial grasses, corn, wheat, and aphid vectors in the disease cycle and epidemiology of BYDV in parts of eastern Canada. Levels of infection, virus variants, patterns of infection in the fields, and development of aphid vectors suggested that local reservoirs such as those in grasses, winter wheat, and corn were of limited importance as virus sources for the crops. Instead, BYDV inoculumn brought in by wind-blown aphids from elsewhere was identified as the primary source of infection in spring cereals. There was no evidence that overwintering aphids play a major role in spreading BYDV from the limited virus source available locally.

Gildow (32) made the interesting observation that *R. padi* and *S. avenae* reared on BYDV-infected oats or senescing tissues were much more likely to mature as winged adults (alatae) than would those reared on healthy plants. Differences in nitrogen metabolism, resulting in increased amino acid concentration in diseased or senescing plants, were considered to be responsible for increased alatae production. If these findings are also true for aphids feeding on other infected cereals, it may play an important role in the secondary spread of the virus within a field because the probability of dispersal of alatae aphids carrying the virus would be greater than the nonwinged aphids (apterae) developing on healthy plants.

The routine use of insecticides on winter cereals to control aphid vectors in order to reduce the spread of BYDV does not appear to be economic in many parts of north America considering the sporadic nature of the virus epidemics and the fact that in some instances the aphids may be virus-free. In other parts of the world, the use of systemic insecticides against vectors has been recommended (1, 95). In any event, development of cereals resistant to the virus and with winter-stress tolerance is the most effective way of managing the losses created by both biotic and abiotic factors. However, it should be noted that breeders lines must be screened against all BYDV variants that prevail in the region because (a) the spectrum and abundance of variants fluctuates from year to year, (b) cereal cultivars react differently to different variants, and (c) breeders lines showing resistance to local variants in one geographical area do not always show resistance in another geographical area (38, 52, 64).

WSSMV has been known to occur only on wheat in Canada, the United States, and in southern France. The actual losses caused by the virus are not known (93).Temperature is the major climatic factor affecting disease development. The wheat sown in infective soil in autumn becomes infected after emergence when soil temperature is around 15°C and the symptoms are not generally evident until growth resumes in spring at temperatures between 5°C and 15°C (92). WSSMV has the lowest temperature range for symptom development known for a plant virus.

Effect of Virus on Winter Cereal Survival

Variations in winter survival may be the prime factor affecting the productivity, and distribution of a crop in northern areas. Freezing stresses on crop plants have been well described over a number of years (60, 96) even though the precise mechanisms of injury, and, in particular, of tolerance remain obscure. However, it is well known now that the site of injury and of tolerance lies in, or is associated with the plasma membrane (96). Plants of many species can withstand intercellular freezing after cold hardening (growth at low temperature), but when intracellular freezing occurs, the affected cells are killed. Depending on the area of the plant frozen, death of the plants, or injury with weak regrowth results after removal of the stress. Minor changes in conditions before the freezing stress may influence the threshold temperature at which such freezing damage may occur. These include the temperature and duration of the low-temperature hardening period, the tissue water content, and various other factors that may interact with expression of the genotype. The genotype sets programmed limits on the levels of freezing tolerance, which are heritable (17).

Factors other than freezing, but related to low temperatures can impinge also on crop plants overwintering in northern areas (77). Flooding can reduce cold-hardiness by raising tissue water content dramatically. Ice encasement of plants may occur after flooding, by snowmelt, or rain during winter, leading to deleterious anaerobic metabolism in the tissues. Snow molds are diseases of cereals and other crops that develop at temperatures around the freezing point and can destroy crop plants in late winter. These factors can influence independently the survival of overwintering cereals, and when they occur in combination, are in most cases even more detrimental to plant survival.

Reduction of the overwintering stand is a frequent result of winter stress and depending on the magnitude of the reduction, it will lead to a decline in the economic yield of the crop. It has been estimated (28) that a reduction of 25% can be compensated for by extra tillering of individual plants leading to maintenance of yield, but reduction below this level will result in yield reduction, to a level dependent on summer growing conditions.

It follows from this brief description of restraints to the survival of the overwintering crop that any extra stress load on the plants before winter will influence winter tolerance, and that this influence is most likely to be detrimental. Infection by a virus may affect the condition of a plant preceding a low-temperature stress event and thus affect development of the plants tolerance to that event, or it may interact directly with the plants ability to survive the stress. Which of these aspects is the most significant in terms of winter survival is not known at this time, but information is available indicating that vector-transmitted virus infections are able to limit winter survival of crop plants and to affect tolerance to individual components of winter stress.

Field Studies

Early reports on the effects of BYDV on winter cereals indicated substantial economic losses due to virus infection, but most effects on wheat were due to growing-period losses and not to winter-stand reduction. However, Endo and Brown (24) reported that conditions that substantially reduced stand survival of barley and oats did not reduce stands of a number of cultivars of wheat, but did induce stunting and reduced a number of yield components. This study indicated that the virus successfully overwintered in relatively severe conditions in which the aphid vectors failed to survive. The aphids used in this study, the apple-grain aphid *Ropalosiphum fitchii* were the most common of a number of species active in the fall. Infections of barley and oats early in the fall gave greater yield loss, greater stunting, and greater winter killing than did infections later in November. The conclusion from these studies over several years was that infections of the young wheat crop occurred from viruliferous aphid vectors from the recently harvested spring oat crop, thus setting a potentially perennial cycle.

In more severe winter conditions, there were significant increases in winterkill of a range of winter wheat cultivars following BYDV infections (27). The vector in these experiments was *R. padi*, the oat-birdcherry aphid. Infections beginning early in fall, 7 days after planting were markedly more severe in terms of winterkill than those begun after 30 days of growth. The early fall infection was also more severe than a spring infection on total grain yield, with reductions of up to 80%. The number of kernels per head and kernel size were reduced markedly by the disease.

In other areas with lower winter stress (2, 95) yield losses were greater from fall virus inoculations than spring inoculations, and in some cases, substantial gains in eventual plant vigor and yield were derived from chemical control of the vector by systemic insecticides. These were particularly effective on the fall infestations. Similarly, in those areas where planting date was flexible, a significant measure of control was obtained by delaying planting until after the autumn flights of the *R. padi* aphids. In contrast, in the state of Washington, observation of severe BYDV infections on spring cereals in close proximity to undamaged winter crops have been explained by the advanced growth stage attained by winter cereals before early summer flights of viruliferous aphids (16). In this study area, aphids occasionally overwintered causing major outbreaks of BYDV and substantial yield losses in the next growing season.

Significant decreases in winter survival in six out of ten cultivars inoculated in the fall with BYDV through *R. padi* aphids were reported in Kansas (69). Stunting of overall growth was observed in all cultivars from fall inoculations, and in most cultivars from spring inoculations. Grain yields were reduced, up to 60% by the fall inoculation and less so, to 30% by the spring inoculation. No cultivars were detected with any tolerance to BYDV in these tests.

A search for BYDV tolerance among 1,200 wheat lines was made by Carrigan et al. (18) after a severe natural infestation of viruliferous *R. padi* aphids in Indiana. An early fall seeding once again was affected much more severely than a late fall seeding when low temperature immobilized the aphid population. A number of lines showed relatively smaller yield decreases in the early BYDV infected planting than the later, less infected planting. The actual loss due to BYDV might well have been greater than the 15% recorded, because an early planting would yield frequently more than a late October planting in the absence of disease. Later years of this study showed considerable BYDV tolerance of two of these lines. No data of winter survival changes were recorded, but general observations were made that stunting, chlorosis, and yield reduction due to BYDV were greater in years of severe winter conditions.

A multiyear study in Illinois, showed similar variation in tolerance to BYDV in a wide array of winter wheat lines, but no major increases in tolerance were found (20). The best performers were crosses with *Agropyron elongatum*. One year of the study with a severe winter showed a 30% decrease in winter survival due to the fall infection, and in other less severe years, fall infected plots were less vigorous after winter but no data on winter survival was presented. In all of the 4 years, fall infections gave rise to much larger yield reductions than the spring infection, and this applied to nearly all the growth parameters and yield components observed. Date to heading was delayed by up to 5 days by fall infection and slightly or not at all by spring infections.

Winter barley is generally less tolerant to winter stress than winter wheat and shows a range of tolerance to BYDV. In Missouri, a series of cultivars and experimental lines of varying tolerance inoculated with strain MO-1 of BYDV from *R. padi* vectors showed variable reductions in winter survival in approximate relation to the susceptibility to the virus (40). Other growth and yield components were reduced also by virus infection. The source of the demonstrated tolerance was not known, but did not include the *Yd2* gene derived from Ethiopian sources. The *Yd2* gene, however, has been incorporated into a winter barley, and the winter survival of the resultant cultivar Vixen was significantly greater than the recipient cultivar Igri in the presence of BYDV infections of varying severity and inoculation times (70). A number of other yield parameters were modified also by the transferred gene in response to BYDV, but all contributed to a substantial increase in grain yield in the presence of the virus. No such resistance gene has been located in wheat: tolerance to the virus in wheat is controlled in a complex manner (19) with relatively low heritabilities.

The basis of the consistent and substantial differences in yield reduction resulting from BYDV infections in the fall as compared to those in the spring is difficult to establish or even rigorously test experimentally. Fall infections due to viruliferous aphids occur on plants in the seedling stage, which is known to be the most susceptible period of the cereal plant to disease development. Weather conditions in the fall are rarely extreme allowing for rapid disease development, such that, by the onset of winter, the vigor and stress tolerance of infected plants is reduced markedly in comparison with noninfected plants. Depending on the severity of the winter, winterkilling and stand reduction may, or may not occur, but winter stresses may reduce further the vigor, the rate of recovery, and regrowth of virus-infected plants after winter. This, in turn, can lead to poor spring growth and eventually the lower yields observed in comparison with spring-infected, or noninfected plants. In contrast, spring infections of BYDV on winter cereals, usually, occur when the plant is very vigorous and in an accelerating growth phase, well past the most susceptible seedling stage, and the disease development period is relatively short. The different effects of the two infection times thus is due to a combination of factors: stage of development of the plant at infection; the interaction of the disease and winter stress; and potentially, the duration of contact between the virus and host plant.

Whereas minor losses due to BYDV infections appear to be frequent in many, if not all cereal growing areas (16, 72, 93), substantial epidemics often are associated with unusual conditions. One such case in eastern Canada has been referred to previously (67). Migratory aphids were deposited on cereal crops during fall and even early winter. Temperatures were considerably higher than normal throughout the fall, and substantial infestations of aphids were observed at many sites in eastern Canada. Populations were mainly of *R. padi*, but with some *R. maidis*, from 15 to 150 aphids per meter of row. Of the field aphids, 69% were BYDV-positive, and there was widespread development of disease symptoms. A number of different virus strains were isolated from the field infections. Development of winter tolerance, as determined by cold-hardiness was found in December to be remarkably low (C.J. Andrews and M.K. Pomeroy, unpublished), which was explained as an association with the warm fall temperatures and BYDV infections, although no experiment had been devised at that time to attempt to separate the two effects. The remainder of the winter was relatively warm, and despite the limited tolerance that plants had developed, winter survival in much of Ontario was relatively high. In contrast, slightly more variable conditions leading to accumulation of water and ice sheeting in some parts of Ontario and much of Quebec led to almost complete kill of winter cereals in those areas due to winter stress in association with BYDV infections. Surviving plants were found to be infected severely with BYDV, thus serving, potentially, to accelerate the cycle of infection. BYDV in parts of Quebec is seen as the most severely limiting factor in the winter survival of cereals, in particular, winter barley. Virus-tolerant lines have been found to have substantially greater winter survival than nontolerant germplasm (A. Comeau, personal communication).

In the above study (67) as in others (16, 24) winter barley, in general, was found to have greater severity of disease symptoms than the winter wheats, and the winter ryes were found to be free of symptoms. Triticales were not symptom-free, but generally were more tolerant than the wheats. Winter oats cannot be grown in this area because of winter stresses, possibly exacerbated by its BYDV susceptibility, but in other areas it shows more severe symptoms than does winter barley. Interpretation of BYDV data in some parts of Ontario was complicated by the presence in winter wheat of advanced symptoms of WSSMV, but the individual effect of symptoms of this virus on survival and plant development could not be determined.

Controlled Environment Studies

In northern latitudes, and to some extent in higher elevations, the extent to which plants can withstand winter is dependent on their ability to survive a number of individual stress components, as briefly discussed above and elsewhere (77). At our Research Centre, we have for the past several years been studying the components of winter stress by the use of controlled environment facilities to determine individual plant responses, and the degree of association between these stress tolerances existing in various genotypes. It was logical from much of the foregoing information that BYDV infections should be considered one of these components of winter stress. There are indeed parallels between BYDV infections and winter stress in general: both the biotic and abiotic stresses are present every year with variable effects on the crop plant, but every few years, much greater effects are seen with highly detrimental effects on crop survival and productivity. We resolved to determine if one stress was predisposing the plant to damage by the other stress, and to determine the magnitude of these interrelationships.

An early logistic difficulty was that environmental requirements for the development of cold tolerance, for aphid feeding, and for virus multiplication are not compatible.Variability during autumn normally provides conditions suitable for inoculation, incubation, and cold-hardening over several months, but such conditions are difficult and inefficient to produce in controlled environments. A successful compromise was achieved with aphid feeding at 15°C day and night for 3 days and with a staged decline in temperature allowing for virus multiplication and disease development leading to a cold-hardening period of 6 to 7 weeks at 2°C-day, 0°C-night temperature, with a 12-h day length. The feeding of nonviruliferous aphids in these experimental conditions induced no measurable effects on wheat or barley plants.

FREEZING TOLERANCE

In initial experiments, a moderately severe strain of BYDV was developed in populations of *Sitobion* (= *Macrosiphum*) *avenae*, the English grain aphid, and placed on 20 to 35-day-old cereal plants that had been held to the 2-leaf stage by low-temperature exposure. Populations of 8 to 10 aphids were distributed on each plant that, at the 15°C-infection temperature and a disease development period of several weeks gave 100% infection of plants as tested by latex agglutination serology (66). Winter cultivars of oats, barley, and wheat in the presence of the virus showed substantially lower levels of cold-hardiness relative to healthy plants at the end of the low-temperature growth period (Table 4.1). Oats and barley showed the greatest loss in cold hardiness, whereas the soft white wheat was more affected than the hard red wheat Kharkov. Puma rye showed only mild symptoms and no loss in hardiness. Differences in symptom expression and

TABLE 4.1. Reductions in cold-hardiness of winter cereals after infection with BYDV or WSSMV prior to 7 weeks of low-temperature growth.*

	$LD_{50}°C^a$			
	BYDV		WSSMV	
	Noninfected	Infected	Noninfected	Infected
Oats-Coastblack	−11.6	− 6.7	—	—
Barley-Dover	−14.7	− 8.5	—	—
Wheat-Fredrick	−17.6	−15.8	−19.6	−15.8
Wheat-Kharkov	−23.2	−22.8	−24.6	−23.4
Rye-Puma	−28.7	−28.8	—	—

* From Ref. 66.
^{a} $LD_{50}°C$ is the temperature at which 50% of the population is killed in a 1°C/h-programmed temperature decline.

cold-hardiness reduction were not due to differential virus multiplication, as all entries showed substantial and approximately similar values in mean virus particle counts by serologically specific electron microscopy (SSEM). Other workers (23) have shown also in winter barley after BYDV infections through two species of aphids a small but significant decrease in tolerance to minor freezing stress (−3.8°C). In these experiments, the

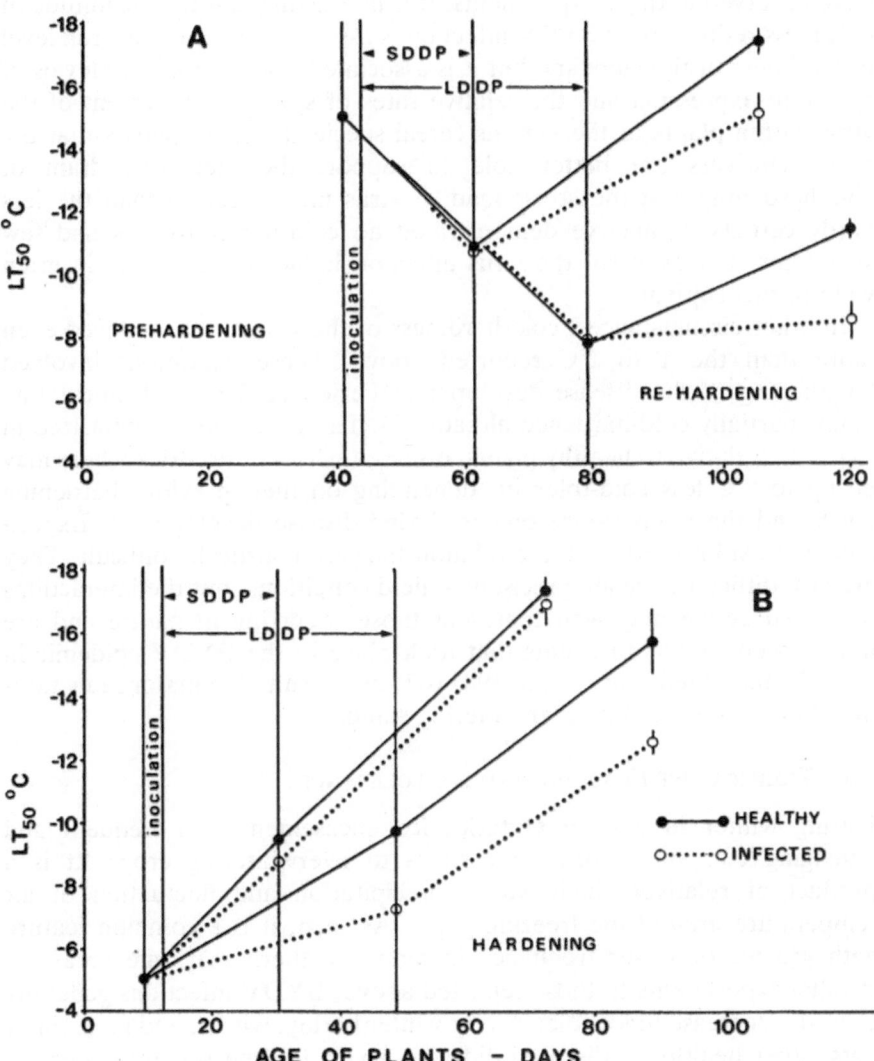

FIGURE 4.1. The cold-hardiness of healthy and BYDV-infected Fredrick winter-wheat with and without cold-hardening before inoculation, and with short (S) and long (L) disease development periods (DDP). From Ref. 3.

addition of inoculations of the fungus *Pyrenophora teres* further reduced the tolerance of crown tissue to freezing.

Cold-hardiness reductions of 5 to 6°C observed (66) in the winter oats and barley are sufficient to reduce winter tolerance to impractical levels in the field, whereas the smaller reductions in the wheats are also of major significance. Fowler et al. (29) have calculated that changes in LD_{50} of 0.5°C will translate into significant changes in overwintering field survival of wheat, and changes due to BYDV substantially greater than that level were observed in these experiments. It is interesting that the magnitude of hardiness reduction by BYDV infection is associated with the overall level of hardiness of the cultivars, but it is associated also with known levels of symptom expression and the relative rates of systemic movement of the virus within plants of the various cereal species (50). It appears that the hardy cultivars are better able to support the metabolic drain of cold-hardening and the stress load of virus multiplication than the less hardy cereals. Puma rye demonstrated no cold-hardiness loss and few symptoms, but even so, the virus infection induced a 14% loss in grain yield in this cultivar.

In other circumstances, cold-hardiness of the wheats was reduced even more than the 1 to 2°C reported above. These conditions involved lengthening of the disease development period (DDP), and inoculating plants partially cold-hardened already (3). The results are summarized in Fig. 4.1. Relative to healthy plants, diseased plants of Fredrick wheat may be up to 8°C less cold-tolerant, depending on their previous hardening levels and the temperature regime during disease development. Experiments to explore further these relationships are logistically difficult. They are also difficult to relate precisely to field conditions, but the interactions described above may well represent those occurring in nature and are likely to be similar to events that took place in the BYDV epidemic in Ontario and Quebec in 1982 to 1983 (67) that resulted in major reductions in cold-hardiness and of overwintering stands.

LOW-TEMPERATURE FLOODING AND ICE TOLERANCE

During winter in eastern Ontario, ice encasement is a frequent and damaging component of winter stress to overwintering crops. It is a product of relatively high winter precipitation and fluctuation of air temperature around the freezing point. As such, it is a common feature with greater or lesser frequency in many northern maritime areas. In parallel experiments to those reported above, BYDV infections generally caused a decrease in ice tolerance of winter barley, wheat, and rye; winter oats either healthy or diseased did not survive ice encasement under the conditions imposed (66) (Table 4.2). Effects on barley were greater than on the wheats. The effect of BYDV on ice tolerance was not however as consistent as that on freezing tolerance. A number of observations were made where infected plants were more ice-tolerant than healthy plants but

TABLE 4.2. Reductions in ice tolerance of winter cereals after infection with BYDV or WSSMV prior to 7 weeks of low-temperature growth.*

	% Survival[a]			
	BYDV		WSSMV	
	Noninfected	Infected	Noninfected	Infected
Barley-Dover	45	5	—	—
Wheat-Fredrick	80	45	65	31
Wheat-Kharkov	100	92	95	90
Rye-Puma	100	100	—	—

* From Ref. 66.
[a] % Survival after 7 days of total ice encasement at −1°C.

TABLE 4.3. Cold-hardiness, ice tolerance, and crown water content of Dover winter barley and Fredrick winter wheat plants with and without BYDV infection and low-temperature flooding, determined at the end of the flooding period.[a]

	$LD_{50}°C^b$		LD_{50} Days[c]		% Crown water content	
	Cold-hardiness		Ice tolerance			
	Nonflood	Flood	Nonflood	Flood	Nonflood	Flood
Barley						
Healthy	−11.2	−4.9	8.0	7.2	77.4	82.0
Diseased	− 9.1	−3.8	7.6	3.1	80.8	84.5
Wheat						
Healthy	−13.8	−7.5	9.0	13.7	76.3	79.8
Diseased	−12.2	−7.2	8.8	9.3	78.0	81.8

[a] From Ref. 4.
[b] $LD_{50}°C$ is the temperature at which 50% of the population is killed in a 1°C/h-programmed temperature decline.
[c] LD_{50} days is duration in which 50% of the population is killed in total ice encasement at −1°C.

no physiological factor was immediately apparent as an explanation. A long-term low-temperature stress also yielded conditions in which infected plants were more tolerant than healthy plants (3). The dry weight of the infected plants was increased leading to the proposal that energy reserves were increased in response to the virus. Jensen (49) reported earlier an increase in carbohydrate in cereals due to infections at warm temperatures but this was later shown to be very unlikely to be associated with increased survival at low temperature [see discussion below on metabolism and Ref. (4)]. No other explanation has emerged yet to explain the rare anomalies. At least one incidence of an increase in field survival by BYDV infection has been reported (27).

Ice encasement is preceded frequently by a period of low-temperature flooding. The flooding may occur also as a relatively independent event in fall or early winter. Low-temperature flooding may be damaging to crops (75, 77) but over longer periods than ice encasement, and with a strong temperature-dependence (15).

TABLE 4.4. Tiller numbers, plant weight, and grain yields of Dover winter barley and Fredrick winter wheat plants with and without BYDV infection and low-temperature flooding.[a]

	Total Tillers no.	Fertile Tillers no.	g/plant	
			Plant weight	Grain weight
Barley, healthy				
Nonflooded	34	20	59.0	19.1
Flooded	37	13	54.8	11.6
Barley, diseased				
Nonflooded	29	13	53.5	9.0
Flooded	17	6	46.2	4.1
Wheat, healthy				
Nonflooded	19	16	72.8	29.7
Flooded	19	15	66.8	28.3
Wheat, diseased				
Nonflooded	24	20	61.7	13.2
Flooded	29	19	65.7	14.5

[a] From Ref. 4.

Flooding during the cold-hardening period reduced the cold-hardiness of winter wheat by 5.3°C and of winter barley by 6.3°C (4) (Table 4.3). Following infection through *R. padi* aphids with a moderately severe vector nonspecific strain of BYDV, the low-temperature growth period induced less cold-hardiness than healthy plants; by 1.6°C in wheat and 2.1°C in barley. The two stresses together (BYDV infection prior to cold-hardening and the 2-week flooding period at the end of the hardening period) gave slightly less than additive effects, bringing hardiness levels, particularly in the barley to very low levels.

Despite this strong effect on cold-hardiness, low-temperature flooding under certain circumstances has the effect of increasing tolerance of cold-hardened wheat to subsequent ice encasement (6, 7). This is an anaerobic hardening response whereby the mild hypoxic stress of flooding increases tolerance to the more severe hypoxia, or anoxia of ice encasement. In this way it is analogous to the process of cold-hardening, where mild low temperatures increase the tolerance of plants to more severe freezing stresses. The introduction of BYDV to the plants prior to the low-temperature growth period removes, or counteracts the promotive effect of flooding on subsequent survival in ice encasement (4) (Table 4.2) in wheat but adds further to the damaging effects of flooding in barley. The metabolic basis of the anaerobic hardening involves an acceleration of glycolysis and an increased supply of cellular ATP (10) but the manner in which it is reversed by the presence of the virus is not yet known.

There are also strong interactions between BYDV infection and low-temperature flooding on subsequent development of winter wheat and barley plants (4) (Table 4.4). Flooding reduced the number of fertile tillers in the barley, particularly in the presence of the virus, whereas there was an increase in fertile tillers due to the virus in wheat. The virus also increased the total number of tillers in wheat, but not in barley. Grain yield per plant was reduced substantially in both cereals in the presence of the virus and there was a near-additive effect of flooding in the barley. Flooding together with BYDV infection caused considerable stunting of barley plants and some produced no inflorescences. The combination of stresses, therefore, appeared to reverse the vernalization process in the barley, but not so in the wheat which has a longer vernalization requirement.

Effect of WSSMV on Cold and Ice Tolerance

The effects of WSSMV on stress tolerances of cereals have not been investigated extensively. One series of experiments was set up as a contrast to BYDV, because WSSMV is a mosaic-type virus that regenerates in mesophyll cells. Although the natural vector of WSSMV is the soil fungus *Polymyxa graminis*, for these experiments the virus was transmitted through sap inoculation using an air brush. WSSMV caused a 3.8°C decline in the cold-hardiness of Fredrick wheat and a 1.2°C decline in the more cold-hardy Karkov wheat (66) (Table 4.1). Similar substantial decreases in ice tolerance of these cereals due to the virus were observed (Table 4.2). Therefore, despite the difference from BYDV in the nature of the disease development, the effect on winter-stress parameters is similar or somewhat greater. Potentially, WSSMV could have dramatic effects on winter-stress tolerance because of its low optimum temperature for disease development.

Metabolic Changes Due to BYDV in Cereals

Barley Yellow Dwarf Virus has been shown to be associated with phloem tissues (25), which explains the manner in which infections become rapidly systemic, and the cause of the disruption of transport of photosynthate in the plant. In severe infections, phloem tissues are destroyed, thus halting transport totally and leading to leaf death. However, many changes can occur before such damage arises, which contribute in various ways to the evident symptoms of the disease generated by BYDV.

Metabolic Effects at Warm Temperatures

Much of the work describing metabolic disturbances induced by BYDV infections has been done by Jensen and his associates. Effects of a severe strain of the virus transmitted by *R. padi* aphids to barley plants induced substantial decreases in chlorophyll, photosynthetic rates, and leaf area, but caused increases in respiration rate on a fresh weight, or leaf-area basis (45). Such increases were not seen on a dry weight basis, because dry weight per unit of fresh weight of leaf increased rapidly with age in infected tissue, whereas the proportion stayed relatively constant in healthy tissue. The decreases in chlorophyll account for the yellows effect of the virus, and the increase in dry weight was assumed to be associated with the reduction in the efficiency of phloem transport such that export from the leaf tissue is retarded.

The development of disease symptoms due to BYDV were greater at 10°C growth temperatures than above (to 27°C) and dry weight accumulation of diseased tissue in relation to the control during growth at 10°C was greater than at higher temperatures. This is, at least partly, a result of a lower respiratory rate on a dry weight basis at 10°C than at higher temperatures (46).

The increase in dry weight due to infection was later shown (47) to be due largely to the accumulation of soluble carbohydrate (up to 15 times that of the content in healthy tissues), a nonsoluble nitrogen component, and a major increase in the small starch component. Soluble proteins and structural carbohydrates were not influenced by the presence of the virus. Symptoms of infection were proposed then to be due to an inhibition of phloem transport out of the leaves, giving rise to major accumulations of sugars, carbohydrates, and nitrogenous compounds, possibly glutamine. As these components are no longer transported to meristematic regions and areas of expansion, the growth rate slows, leading to overall plant stunting. High levels of a number of sugars are likely to inhibit photosynthetic reactions (13) that together with severe reductions in total chlorophyll, can account for observed reductions in photosynthetic rates due to infections. A number of studies on other host–virus combinations, reviewed by Fraser (30) have concluded that lowered photosynthetic rates following virus infection are associated frequently with physical damage to the chloroplast structure, the deterioration of its membranes, and the deposition of large starch grains within the chloroplast envelope.

The foregoing observations by Jensen were made on spring barley plants, and similar observations were later made on a series of spring wheat cultivars. Massive accumulations of soluble carbohydrates were observed as a result of BYDV infection from *R. padi* vectors particularly in the laminae of the upper leaves, whereas the culm and lower leaf sheaths showed little or no increase (49). Increases in carbohydrate in the upper parts of the plant were greatest in a BYDV field-susceptible cultivar and least in a resistant cultivar. Whereas respiration was increased markedly by the virus in the susceptible cultivar, it was depressed by BYDV in the tolerant cultivar (51). Reports of such changes in winter wheat have not been found, but it is assumed that if such major changes occur in the various species of the spring cereals they are likely also to occur in the winter types.

Metabolic Effects at Low Temperatures

The loss in cold-hardiness brought about by the presence of BYDV infections in cereal plants is large enough to be significant in the field survival of overwintering crops. In order to understand better the mechanisms by which cold-hardiness is reduced, a series of experiments were undertaken, initially with two winter wheats of differing susceptibility to the virus, and two BYDV strains of differing severity. The more tolerant cultivar was Caldwell, a hard, red winter wheat, and the susceptible type was Fredrick, a soft, white winter wheat, generally also less cold-tolerant than the red wheat. The BYDV strains were RMV, a mild strain specifically transmitted by the vector *Ropalosiphum maydis* and PAV, a severe strain specifically transmitted by both *R. padi* and *Sitobion avenae*.

The tolerant cultivar Caldwell was reduced in its cold-hardiness equally by both strains of the virus, but only by about 1°C. The more susceptible Fredrick was reduced in cold-hardiness more by the severe PAV strain than by the RMV strain and the reduction was twice that observed in the more tolerant Caldwell (Table 4.5). Total nonstructural carbohydrates in the crown increased during the incubation period, and in the cold-hardening period, but the increase was less in the presence of the severe

TABLE 4.5. Changes in a number of physiological parameters in 2 winter wheat cultivars, Caldwell (tolerant), and Fredrick (susceptible).

| | | | | Moisture | Respiration | Carbohydrate | |
| | | | | | | Crown | Leaf |
		LD_{50}°C	Fresh wt.gm	(%)	(O_2 nmol/g/min)	μm/g dry wt.	
Caldwell							
H[a]	O[b]	—	2.60	90.80	5.44	786	658
	PI	−7.0	8.89	87.95	3.15	1287	844
	PH	−18.8	11.43	77.70	1.70	2706	2060
RM	O	—	2.60	90.80	5.44	786	658
	PI	−6.8	7.20	88.03	3.16	1200	1010
	PH	−17.8	9.65	76.10	1.85	2876	2677
PAV	O	—	2.60	90.80	5.44	786	685
	PI	−6.8	5.24	87.58	3.61	1052	930
	PH	−17.9	11.65	76.93	2.30	2264	2494
Frederick							
H	O	—	3.16	88.98	4.35	854	700
	PI	−7.0	10.83	86.10	2.79	1358	1182
	PH	−16.5	16.82	77.10	1.58	2612	2158
RM	O	—	3.16	88.98	4.35	854	700
	PI	−5.9	8.83	86.73	3.27	1217	1140
	PH	−14.6	12.48	76.85	1.48	2666	2339
PAV	O	—	3.16	88.98	4.35	854	700
	PI	−6.0	8.11	84.76	2.84	1604	1540
	PH	−14.0	13.04	77.63	1.79	2515	2280

[a] H = healthy; RM = infected with a mild strain of BYDV; PAV = infected with a severe strain of BYDV.
[b] O = sampled before infection; PI = sampled after infection; PH = sampled after coldhardening.

strain of the virus particularly in the more tolerant cultivar. In the leaves, similar increases occured in healthy tissues, and both strains of the virus stimulated marginally increased levels of carbohydrate in both cultivars.

This was a very different result than the massive increases reported by Jensen at warmer growing temperatures, and it reflects very different carbohydrate metabolism occurring at the lower temperatures. This can allow the accumulation of up to 55% dry weight of carbohydrate in crowns and leaf bases in various conditions of low temperature (11). Small increases did occur in response to the virus in leaf tissues, close to the photosynthetic source and in accord with Jensen's observations, but the decreases observed in the crown may well be associated with the general decline in cold-hardiness.

Respiration generally declined by about 60% during the incubation period and a further 30% after low-temperature hardening. After hardening in the presence of the virus, plant respiration was greater in response to the severe than to the mild strain, particularly in the tolerant cultivar (Table 4.5). These results again were in marked contrast to the observations of Jensen, but it is clear that the much higher overall levels of carbohydrate at low temperature make comparisons between the two temperature regimes of little value. It is possible that the increased level of respiration in the presence of the severe strain of the virus is associated with the decreased level of crown carbohydrate after the low-temperature period, whereas the mild strain has little or no effect on respiration or crown carbohydrate. As both virus strains contribute to significant reductions in cold-hardiness, factors in addition to carbohydrate reserves and respiratory potential must mediate effects of the virus.

A very common feature of virus-infected cereal plants growing at low temperature is an increase in moisture content of the crown and, to a lesser extent, of the leaves (3, 4) together with a general decrease in the weight of the plants (Table 4.5, 4.6). Changes in moisture content are correlated closely with changes in cold-hardiness and have been suggested as a

TABLE 4.6. Changes in water content and carbohydrate fractions in infected and non-infected Frederick wheat plants.[a]

	Plant	Water Content (%)		TNSC[b] (mmol/g dry wt)		Fructan	
	Fresh wt (g)	Crown	Leaf	Crown	Leaf	Crown	Leaf
After infection							
Healthy	20.85	80.9	82.5	2.29	1.44	0.59	0.05
Diseased	14.93	83.1	78.2	1.70	1.99	0.42	0.28
After hardening							
Healthy	28.15	66.5	69.7	2.77	2.77	1.70	1.80
Diseased	19.37	71.8	70.6	2.36	2.61	1.07	1.22

[a] Plants infected with a severe strain of BYDV; measurements made after a disease development period and coldhardening.
[b] Total nonstructural carbohydrates.

predictive test as the correlations are higher than with any other of over 30 factors measured during low-temperature growth (29). Similar high correlations have been found with BYDV-infected and noninfected cereals, but in a number of unusual cases in which severe hardiness reductions were recorded, lower than predicted moisture contents were seen that reduced the correlations (3). Changes in moisture content are thus symptoms of virus infection over a relatively broad range of disease severity, but the relationship appears to be obscure. It seems unlikely that such changes will prove useful to explain the association between BYDV infections and cold-hardiness reduction.

A number of other metabolic changes in cold-hardened plants due to BYDV infections have been determined in Fredrick wheat with the PAV strain transmitted to individually caged plants by R. padi vectors. In this series of experiments, total nonstructural carbohydrates were reduced in the crowns, and marginally so in the leaves (Table 4.6). A number of other carbohydrate components were reduced by the disease, in particular the fructan component that declined by 30 to 40% in crowns and leaves after low-temperature growth in the presence of the virus.

The decline in fructans is particularly pertinent to changes in cold-hardiness of the tissues. Suzuki and Nass (97) have described recently a close correlation between levels of a relatively long-chain fructan component in crown tissue of low-temperature grown wheat and rye and their cold-hardiness. The mechanistic basis of the correlation was not established but a number of proposals for the cryoprotective effect of fructan were put forward. These include its lower osmotic activity than free sugars, allowing for greater dehydration of tissues, its lack of penetrative crystal structure, and the low temperature optima of its metabolic enzymes, allowing for accumulation and metabolism of the reserve materials below the freezing point. The degree of polymerization of the fructan influenced by BYDV infection in these experiments is not known yet, neither is the mechanism known by which fructan is reduced by the presence of the virus.

Carbohydrate levels and changes in respiration discussed above can influence the overall metabolism of the plants by maintaining energy supply, particularly during stress periods, in the form of ATP. Supply of this adenylate is required for many maintenance processes in cells, the operation of plasma-membrane transport enzymes for nutrient uptake and maintenance of cell turgor, and in virus-infected tissue, the viral-directed synthesis of virus from cellular components. Measurable changes in ATP levels are frequently small; crowns of Fredrick winter wheat infected with the severe PAV strain of BYDV from R. padi vectors showed significant changes in ATP after incubation and after cold-hardening, with minor changes in total adenylates (Table 4.7). Changes in adenylate energy charge (AEC), a ratio of ATP to the total adenylate pool (12) were not seen, or showed small variations only. Changes in the leaves were smaller than those in the crown.

TABLE 4.7. Changes in adenylates and [86]rubidum uptake into crowns and leaves of infected and noninfected Fredrick wheat.[a]

	AdN[b]	ATP	AEC[c]	[86]Rb uptake
	(nmol/mg protein)			(dpm/mg dry wt.)
After infection				
Crown Healthy	23.1	14.5	0.77	2054
Crown Diseased	23.1	16.5	0.84	2584
Leaf Healthy	8.0	4.7	0.79	256
Leaf Diseased	7.6	4.5	0.82	215
After hardening				
Crown Healthy	13.9	8.0	0.74	774
Crown Diseased	14.6	9.1	0.70	1497
Leaf Healthy	7.4	4.2	0.71	30
Leaf Diseased	8.2	4.3	0.71	89

[a] Plants infected with a severe strain of BYDV; measurements made after a disease development period and after coldhardening.
[b] Total adenine nucleotides (ATP + ADP + AMP).
[c] Adenylate energy charge.

Crown tissues of low-temperature-grown Fredrick wheat also accumulated more radioactive label when infected with BYDV than when healthy (Table 4.7). [86]Rubidium is a close analog of potassium which is in continual flux across the plasma membrane via ATP-dependent transport enzymes in response to environmental changes. Previously, it has been shown that the ability of wheat crowns to take up rubidium declines with prolonged damaging low-temperature stress (31), yet is maintained in crown tissues that survive a freezing stress that kills contiguous root tissue (77). After a disease development period, the uptake capacity of crown tissue for rubidium was increased in the presence of BYDV (Table 4.7). Further low-temperature growth generally decreased uptake, but the presence of the virus in crowns allowed a doubling of rubidium uptake relative to healthy tissue, and an even greater relative increase in the leaves. The increase in this active uptake is assumed to be due to an effect on plasma membrane ATPases (78). An earlier report (53) from a study of Tobacco Mosaic Virus in tobacco leaves at warm temperatures described a rapid increase in the activity of membrane-bound ATPase after infection and indicated that the increase is found in tissue adjacent to, but not infected by the virus yet. It is tempting to propose that such an increase in membrane transport activity is related to increases in cellular ATP, as at least 25% of cellular ATP is utilized in plasma membrane transport processes (71). As yet such associations involving BYDV infections are unsubstantiated.

The observed increase in ATP and membrane transport activity is further evidence of a general increase in metabolic activity as a response to viral infection. It is uncertain whether the increase is directed by the virus, or whether the response is a defense mechanism by the host tissue. The origin of such a metabolic change due to virus infection has been acknowledged as being very difficult to determine (30, 54). Increased respiratory rates and carbohydrate losses due to the disease at low temperature in the more BYDV-tolerant cultivar Caldwell than in the more susceptible Fredrick would indicate a defense mechanism, but more evidence is required to establish such a conclusion.

Metabolic Effects During Low-Temperature Flooding and Ice Encasement

Changes in a number of metabolic components in crowns and leaf bases of wheat and barley were followed after flooding and during ice encasement, to assess the influence of BYDV infections in the tissue and its effects on the interactions previously described between these two stresses. Carbohydrate in the crown tissue was reduced by a 2-week flooding exposure, and was similarly or more reduced by the presence of the virus (Table 4.8).

TABLE 4.8. Total nonstructural carbohydrates (TNSC) in crown and leaf base tissues of Dover winter barley and Fredrick winter wheat, with and without BYDV infection and low-temperature flooding.[a]

	TNSC (mmol/g dry wt.)			
	Crown	Leaf base	Leachate	Utilized
Barley, healthy				
Nonflooded	2.99	3.64		
Loss in ice	0.55	0.62	0.35[b]	0.82[c]
Flooded	2.70	3.36		
Loss in ice	0.42	0.57	0.40	0.59
Barley, diseased				
Nonflooded	2.17	3.06		
Loss in ice	0.49	0.84	0.55	0.78
Flooded	1.51	2.49		
Loss in ice	0.48	0.71	0.29	0.90
Wheat, healthy				
Nonflooded	2.44	3.82		
Loss in ice	0.44	0.83	0.16	1.11
Flooded	1.86	3.31		
Loss in ice	0.38	0.71	0.16	0.93
Wheat, diseased				
Nonflooded	1.89	3.30		
Loss in ice	0.54	1.06	0.31	1.29
Flooded	1.70	2.93		
Loss in ice	0.43	0.80	0.28	0.95

[a] From Ref. 4. Determinations at the end of the flooding period and after 14 days ice encasement.
[b] Leachate is carbohydrate lost from the tissues (crown and leaf base combined) during ice encasement and thawing, but not consumed by metabolic functions.
[c] Combined loss, during ice encasement of TNSC from crown and leaf base, minus the leachate thus representing TNSC utilized by the tissues.

Flooding in the presence of the virus reduced the content by 50%. Leaf bases showed similar changes but with lower amplitude, presumably reflecting their closer proximity to the photosynthetic source, and the shorter path through damaged phloem tissue. The further loss of carbohydrate in the crown in 14 days of ice encasement following flooding, or without previous flooding was approximately similar in all cases, with or without the virus, and appeared to be independent of the level of carbohydrate prior to ice exposure. The measurement of the utilization of carbohydrate was also approximately similar in all cases and confirmed earlier statements (58, 76) that carbohydrate supply is not limiting to survival during ice encasement.

The loss of carbohydrate during ice encasement is accompanied by the accumulation of anaerobic metabolites in an approximately stoichiometric relationship (8). The accumulation of ethanol in crown tissues during ice encasement was increased by a previous flooding treatment and by the presence of the virus, and increased near additively by both stresses (Table 4.9). CO_2 accumulation, however was decreased by previous flooding, and BYDV has a small effect only in combination with the flooding stress. It is possible that the changes in CO_2 levels are a significant part of the hardening interaction between flooding and ice encasement, and its

TABLE 4.9. Accumulation of CO_2 and ethanol in crown tissue of Dover winter barley and Fredrick winter wheat during ice encasement, with and without BYDV infection and low-temperature flooding.[a]

	Ice period (days)	CO_2		Ethanol	
		Healthy	Diseased	Healthy	Diseased
		(μmols/g dry wt.)			
Barley					
Nonflooded	0	5	8	0	0
	4	117	160	291	297
	9	188	197	360	543
	14	405	336	497	636
Flooded	0	6	10	57	34
	4	128	167	244	448
	9	197	250	519	758
	14	347	432	645	871
Wheat					
Nonflooded	0	8	6	0	1
	4	100	153	211	193
	9	230	234	322	356
	14	387	401	335	580
Flooded	0	9	8	14	50
	4	122	157	213	320
	9	208	215	428	562
	14	347	403	524	647

[a] From Ref. 4.

disturbance by BYDV described in the section on controlled environment studies. Carbon dioxide accumulates to toxic levels during ice encasement (5, 79) and there is recent evidence that CO_2 and its related ions, HCO_3^- and CO_3^{2-} inhibit plasma membrane ATPase activity in a very similar way that ice encasement itself inhibits such ATPase activity (9).

It is not clear in what way BYDV infection stimulates the production of ethanol and, to a lesser extent, CO_2 during ice encasement, particularly as the rate of utilization of carbohydrate apparently is not stimulated by the virus. The virus contributes to carbohydrate utilization before ice exposure, which is associated with an increased aerobic respiratory rate in nonflooded plants and an increased anaerobic pathway in flooded plants. During ice encasement, it is possible that accumulated intermediates from such stimulated pathways are reduced to ethanol without further acceleration of carbohydrate breakdown. Such intermediates are not known but could involve pyruvate or malate (55).

Conclusions

In the few decades since the original description of barley yellow dwarf virus as a pathogenic entity, a large amount of information has been reported on the distribution of the virus, its epidemiology in various areas, vector relationships, and tolerance to the virus. This is strong indication of the general importance that the disease associated with this virus has been given as a result of yield reduction and economic destabilization. The disease fluctuates year to year in its severity. Some of this fluctuation has been explained by aphid vector movements that can be monitored on a relatively routine basis, and that are very much controlled by prevailing weather and wind conditions. In this chapter, we have concentrated on the effects of BYDV and, to a much lesser extent, on WSSMV on the tolerance of cereals to natural and simulated winter stresses, and the interactions of BYDV with a number of these isolated stresses.

Infection of cereals with both BYDV and WSSMV in nearly all cases reduces winter survival, and reduces tolerance to freezing and to anaerobic stresses. The reductions in these tolerances are associated with an increase in respiratory activity, and in the appropriate conditions by increases in fermentative pathways. The precise manner in which these changes are associated is not yet clear. A number of possible metabolic links are proposed, but few of the definitive experiments required to establish discrete pathways have been completed. Relatively little work on the metabolic effects of the presence of the virus at its optimum condition of symptom development has been reported. A recent review of the biochemistry of virus-infected plants (30) listed few studies on BYDV but indicated that research efforts had been spread over the effects of many viruses. This is surprising in the light of the world-wide effects of BYDV disease and its overall impact on agricultural productivity. This lack of basic information makes interpretation of metabolic events involved in stress interactions such as these described here, even more difficult, as the framework with which to compare metabolism perturbed by the various stresses is lacking. It is possible, for example, that in cereal plants infected with BYDV there are modifications in membrane fluidity and permeability similar to those reported in virus-infected tobacco plants (91) and in mammalian cells (26). Such changes could well be associated with reductions in growth rates at warm temperatures, and interference with cold-hardening processes at low temperatures.

There is a strong need for information on the effects of BYDV infection on various aspects of plant metabolism—those that are known to be influenced already, and those that will emerge undoubtedly as being affected by the virus. Such studies must lead eventually to information on the molecular control of virus multiplication. It is apparent that tolerance to the virus is not a simple reaction, particularly in wheat. Knowledge of the details of the perturbed metabolism will identify those aspects of the metabolism most influenced by the virus. Changes to these biochemical events are, therefore, the most likely to yield incremental increases in tolerance to the virus after genetic manipulations through molecular biology and plant breeding.

Acknowledgments. This chapter is dedicated to the memory of Dr. Yogesh C. Paliwal, who died suddenly on June 23, 1988.

References

1. A'Brook, J., 1974, Barley yellow dwarf virus: What sort of a problem? *Ann. Appl. Biol.* **77**:92–96.
2. A'Brook, J. and Dewar, A.M., 1980, Barley yellow dwarf virus infectivity of alate aphid vectors in west Wales, *Ann. Appl. Biol.* **96**:51–58.
3. Andrews, C.J. and Paliwal, Y.C., 1983, The influence of preinfection cold hardening and disease development period on the interaction between barley yellow dwarf virus and cold stress tolerance in wheat, *Can. J. Bot.* **61**:1935–1940.
4. Andrews, C.J. and Paliwal, Y.C., 1986, Effects of barley yellow dwarf virus and low temperature flooding on cold stress tolerance of winter cereals, *Can. J. Plant Pathol.* **8**:311–316.
5. Andrews, C.J. and Pomeroy, M.K., 1979, Toxicity of anaerobic metabolites accumulating in winter wheat seedlings during ice encasement, *Plant Physiol.* **64**:120–125.
6. Andrews, C.J. and Pomeroy, M.K., 1981, The effect of flooding pretreatment on cold hardiness and survival of winter cereals in ice encasement, *Can. J. Plant Sci.* **61**:507–513.
7. Andrews, C.J. and Pomeroy, M.K., 1983, The influence of flooding pretreat-

ment on metabolic changes in winter cereal seedlings during ice encasement, *Can. J. Bot.* **61**:142–147.

8. Andrews, C.J. and Pomeroy, M.K., 1989a, Physiological properties of plants affecting ice encasement tolerance, *Icel. Agric. Sci.* **2**:41–51.

9. Andrews, C.J. and Pomeroy, M.K., 1989b, Ice encasement injury at the cellular and membrane level, *Icel. Agric. Sci.* **2**:57–61.

10. Andrews, C.J. and Pomeroy, M.K., 1989c, Metabolic acclimation to hypoxia in winter cereals; low temperature flooding increases adenylates and survival in ice encasement, *Plant Physiol.* **91**:1063–1068.

11. Andrews, C.J., Seaman, W.L., and Pomeroy, M.K., 1984, Changes in cold hardiness ice tolerance and total carbohydrates of winter wheat under various cutting regimes, *Can. J. Plant Sci.* **64**:547–558.

12. Atkinson, D.E., 1968, The energy charge of the adenylate pool as a regulatory parameter. Interaction with feedback modifiers, *Biochemistry* **7**:4030–4034.

13. Azcon-Bieto, J., 1983, Inhibition of photosynthesis by carbohydrates in wheat leaves, *Plant Physiol.* **73**:681–686.

14. Barr, D.J. and Slykhuis, J.T., 1969, Zoosporic fungi associated with wheat spindle streak mosaic in Ontario. *Can. Plant Dis. Surv.* **49**:112–113.

15. Beard, J.B. and Martin, D.P., 1970, Influence of water temperature on submersion tolerance of four grasses, *Agron. J.* **62**:257–259.

16. Bruehl, G.W., McKinney, H.H., and Toko, H.V., 1959, Cereal yellow dwarf as an economic factor in small grain production in Washington 1955–58, *Plant Dis. Reptr.* **43**:471–474.

17. Brule-Babel, A.L. and Fowler, D.B., 1988, Genetic control of cold hardiness and vernalization requirement in winter wheat, *Crop Sci.* **28**:879–884.

18. Carrigan, L.L., Ohm, M.W., Foster, J.E., and Patterson, F.L., 1981, Response of winter wheat cultivars to barley yellow dwarf infection, *Crop Sci.* **21**:377–380.

19. Cisar, G., Brown, C.M., and Jedlinski, H., 1982, Diallel analyses for tolerance in winter wheat to the barley yellow dwarf virus, *Crop Sci.* **22**:328–333.

20. Cisar, G., Brown, C.M., and Jedlinski, H., 1982, Effect of fall or spring infection and sources of tolerance of barley yellow dwarf of winter wheat, *Crop Sci.* **22**:474–478.

21. Comeau, A. and Dubuc, J.P., 1977, Observations on the 1976 barley yellow dwarf epidemic in eastern Canada, *Can. Plant Dis. Surv.* **57**:42–44.

22. Dedryver, C.A. and Robert, Y., 1981, Ecological role of maize and cereal volunteers as reservoirs for gramineae virus transmitting aphids. *Proceedings 3rd Conference on Virus Diseases of Gramineae*, Harpenden, Rothamsted Experimental Station, pp. 61–66.

23. Delserone, L.M., Cole, H., and Frank, J.A., 1987, The effects of infections by *Pyrenophora teres* and barley yellow dwarf virus on the freezing hardiness of winter barley, *Phytopathology* **77**:1435–1437.

24. Endo, R.M. and Brown, C.M., 1962, Survival and yield of winter cereals affected by yellow dwarf, *Phytopathology.* **52**:624–627.

25. Esau, K., 1957, Phloem degeneration in gramineae affected by the barley yellow dwarf virus, *Amer. J. Bot.* **44**:245–251.

26. Fernandez-Puentes, C. and Carrasco, L., 1980, Viral infection permeabilizes mammalian cells to protein toxins, *Cell* **20**:769–775.

27. Fitzgerald, P.J. and Stoner, W.N., 1967, Barley yellow dwarf studies in wheat.

I. Yield and quality of hard red winter wheat infected with barley yellow dwarf virus, *Crop Sci.* **7**:337–341.

28. Fowler, D.B., Gusta, L.V., Bowren, K.E., Crowle, W.L., Mallough, E.D., McBean, D.S., and Melver, R.N., 1976, Potential for winter wheat production in Saskatchewan, *Can. J. Plant Sci.* **56**:455–50.

29. Fowler, D.B., Gusta, L.V., and Tyler, N.J., 1981, Selection for winter hardiness in wheat III. Screening methods, *Crop Sci.* **21**:896–901.

30. Fraser, R.S.S. 1987, Biochemistry of virus-infected plants. Letchworth U.K., Research Studies Press, p. 259.

31. Gao, J-Y., Andrews, C.J., and Pomeroy, M.K., 1983, Interactions among flooding, freezing and ice encasement in winter wheat, *Plant Physiol.* **72**:303–307.

32. Gildow, F.E., 1980, Increased production of alatae by aphids reared on oats infected with barley yellow dwarf virus, *Annu. Entomol. Soc. Am.* **73**:343–347.

33. Gildow, F.E., 1987, Virus-membrane interactions involved in circulative transmission of Luteoviruses by aphids, (ed) Harris, K.F. in *Current Topics in Vector Research* Vol. 4, Springer-Verlag, New York/London, pp. 93–120.

34. Gildow, F.E., Ballinger, M.E., and Rochow, W.F., 1983, Identification of double-stranded RNAs associated with barley yellow dwarf virus infection of oats, *Phytopathology* **73**:1570–1572.

35. Gill, C.C., 1967, Transmission of barley yellow dwarf virus isolates from Manitoba by five species of aphids, *Phytopathology* **57**:713–718.

36. Gill, C.C., 1970, Epidemiology of barley yellow dwarf in Manitoba and effect of the virus on yield of cereals, *Phytopathology* **60**:1826–1830.

37. Gill, C.C., 1980, Assessment of losses on spring wheat naturally infected with barley yellow dwarf virus, *Plant Disease* **64**:197–203.

38. Gill, C.C. and Buchannon, K.W., 1972. Reaction of barley hybrids from crosses with C.I. 5791 to four isolates of barley yellow dwarf virus, *Can. J. Plant Sci.* **52**:305–309.

39. Gill, C.C. and Chong, J., 1979, Cytopathological evidence for the division of barley yellow dwarf virus isolates into two subgroups, *Virology* **95**:59–69.

40. Grafton, K.F., Poehlman, J.M., Sechler, D.T., and Sehgal, O.P., 1982, Effect of barley yellow dwarf virus infection on winter survival and other agronomic traits in barley, *Crop Sci.* **22**:596–600.

41. Harris, K., 1979, Leafhoppers and aphids as biological vectors: Vector-virus relationships. (eds): Maramorosh, K. and Harris, K.F. in *Leafhopper Vectors and Plant Disease Agents*, Academic Press, New York, pp. 217–308.

42. Harris, K.F., Bath, J.E., Thottappilly, G., and Hooper, G.R., 1975, Fate of pea enation mosaic virus in PEMV-infected pea aphids, *Virology* **65**:148–162.

43. Harrison, B.D. and Murant, A.F., 1984, Involvement of virus-coded proteins in transmission of plant viruses by vectors. (eds): Mayo, M.A. and Harrap, K.A. in *Vectors in Virus Biology*, Academic Press, London, pp. 1–36.

44. Hooper, G.R. and Wiese, M.V., 1972, Cytoplasmic inclusions in wheat affected by wheat spindle streak mosaic, *Virology* **47**:664–672.

45. Jensen, S.G., 1968a, Photosynthesis, respiration and other physiological relationships in barley infected with Barley Yellow Dwarf Virus, *Phytopathology* **58**:204–208.

46. Jensen, S.G., 1968b, Factors affecting respiration in barley yellow dwarf virus-infected barley, *Phytopathology* **58**:438–443.
47. Jensen, S.G., 1969, Composition and metabolism of barley leaves infected with barley yellow dwarf virus, *Phytopathology* **59**:1694–1698.
48. Jensen, S.G., 1969, Occurrence of virus particles in the phloem tissue of BYDV-infected barley, *Virology* **38**:83–91.
49. Jensen, S.G., 1972, Metabolism and carbohydrate composition in barley yellow dwarf virus-infected wheat, *Phytopathology* **62**:587–592.
50. Jensen, S.G., 1973, Systemic movement of barley yellow dwarf virus in small grains, *Phytopathology* **63**:854–856.
51. Jensen, S.G. and Van Sambeek, J.W., 1972, Differential effects of barley yellow dwarf virus on the physiology of tissues of hard red spring wheat, *Phytopathology* **62**:290–293.
52. Jones, A.T. and Catherall, P.L., 1970, The effect of different virus isolates on the expression of tolerance to barley yellow dwarf virus in barley, *Ann. Appl. Biol.* **65**:147–152.
53. Kasamo, K. and Shimomura, T., 1978, Response of Membrane bound Mg^{2+} activated ATPase of tobacco leaves to Tobacco Mosaic Virus, *Plant Physiol.* **62**:731–734.
54. Kassanis, B., 1981, Some speculations on the nature of the natural defence mechanisms of plants against virus infections, *Phytopath. Z.* **102**:277–291.
55. Lance, C. and Rustin, P., 1984, The central role of malate in plant metabolism, *Physiol. Veg.* **22**:625–641.
56. Langenberg, W.G. and Schroeder, H.F., 1973, Endoplasmic reticulum—derived pinwheels in wheat infected with wheat spindle streak mosaic virus, *Virology* **55**:218–223.
57. Lister, R.M. and Rochow, W.F., 1979, Detection of barley yellow dwarf virus by enzyme-linked immunosorbent assay, *Phytopathology* **69**:649–654.
58. McKersie, B.D., McDermott, B.M., Hunt, L.A., and Poysa, V., 1982, Changes in carbohydrate levels during ice encasement and flooding of winter cereals, *Can. J. Bot.* **60**:1822–1826.
59. Mueller, W.C. and Rochow, W.F., 1961, An aphid-injection method for the transmission of barley yellow dwarf virus, *Virology* **14**:253–258.
60. Olien, C.R., 1964, Freezing processes in the crown of Hudson barley, *Crop Sci.* **4**:91–95.
61. Paliwal, Y.C., 1977, Rapid diagnosis of barley yellow dwarf virus in plants using serologically specific election microscopy, *Phytopath. Z.* **89**:25–36.
62. Paliwal, Y.C., 1979, Serological relationships of barley yellow dwarf virus isolates, *Phytopath. Z.* **94**:8–15.
63. Paliwal, Y.C., 1982, Identification and annual variation of variants of barley yellow dwarf virus in Ontario and Quebec, *Can. J. Plant Pathol.* **4**:59–64.
64. Paliwal, Y.C., 1982, Detection of barley yellow dwarf virus in aphids by serologically specific electron microscopy, *Can. J. Bot.* **60**:179–185.
65. Paliwal, Y.C., 1982, Role of perennial grasses, winter wheat, and aphid vectors in the disease cycle and epidemiology of barley yellow dwarf virus, *Can. J. Plant Pathol.* **4**:367–374.
66. Paliwal, Y.C. and Andrews, C.J., 1979, Effects of barley yellow dwarf and wheat spindle streak mosaic viruses on cold hardiness of cereals, *Can. J. Plant Pathol.* **1**:71–75.

67. Paliwal, Y.C. and Comeau, A., 1984, Epidemiology of barley yellow dwarf virus in Ontario and Quebec in 1982 and 1983, *Can. Plant Dis. Surv.* **64**:21–23.
68. Paliwal, Y.C. and Sinha, R.C., 1970, On the mechanism of persistence and distribution of barley yellow dwarf virus in an aphid vector, *Virology* **42**:668–680.
69. Palmer, L.T. and Sill, W.H., 1966, Effect of barley yellow dwarf virus on wheat in Kansas, *Plant Dis. Reptr.* **50**:234–238.
70. Parry, A.L. and Habgood, R.M., 1986, Field assessment of the effectiveness of a barley yellow dwarf virus resistance gene following its transference from spring to winter barley, *Ann. Appl. Biol.* **108**:395–401.
71. Penning de Vries, F.W.T., 1975, The cost of maintenance processes in plant cells, *Ann. Bot.* **39**:77–92.
72. Plumb, R.T., 1983, Barley yellow dwarf virus—a global problem, (eds): Plumb, R.T., Thresh, J.M. in Plant Virus Epidemiology, Blackwell Scientific, Oxford/London/Boston, pp. 185–198.
73. Plumb, R.T., 1974, Properties and isolates of barley yellow dwarf virus, *Ann. Appl. Biol.* **77**:87–91.
74. Plumb, R.T., Lennon, E., 1981, Serological diagnosis of barley yellow dwarf virus (BYDV). Report of Rothamsted Experimental Station for 1980. Part 1:181.
75. Pomeroy, M.K. and Andrews, C.J., 1979, Metabolic and ultrastructural changes associated with flooding at low temperature in winter wheat and barley, *Plant. Physiol.* **64**:635–639.
76. Pomeroy, M.K. and Andrews, C.J., 1983, Responses of winter cereals to various low temperature stresses. (eds): Randall, D.D., Blevins, G., Larson, R.L. and Rapp, B.J. in Current Topics in Plant Biochemistry and Physiology, Vol. 2. University of Missouri, Columbia, pp. 96–106.
77. Pomeroy, M.K. and Andrews, C.J. 1989. Low temperature injury in winter cereals. (ed): Li, P.H. in Low Temperature Stress Physiology in Crops, CRC Press, Boca Raton, Fla., pp. 107–122.
78. Pomeroy, M.K., Pihakaski, S.J. and Andrews, C.J., 1983, Membrane properties of isolated winter wheat cells in relation to icing stress, *Plant Physiol.* **72**:535–539.
79. Rakitina, Z.G., 1971, Effect of an ice crust on gas composition of the internal atmosphere in winter wheat, *Sov. Plant Physiol.* **17**:907–912.
80. Rochow, W.F., 1969, Specificity in aphid transmission of a circulative plant virus. (ed): Maramorosch, K. in *Viruses, Vectors, and Vegetation*, Interscience, New York/London/Sydney/Toronto, pp. 175–198.
81. Rochow, W.F., 1970, Barley yellow dwarf virus: phenotypic mixing and vector specificity, *Science* **167**:875–878.
82. Rochow, W.F., 1972, The role of mixed infections in the transmission of plant viruses by aphids, *Annu. Rev. of Phytopathol.* **10**:101–124.
83. Rochow, W.F., 1973, Selective virus transmission of *Rhopalosiphum padi* exposed sequentially to two barley yellow dwarf viruses, *Phytopathology* **63**:1317–1322.
84. Rochow, W.F., 1977, Dependent virus transmission from mixed infections, (eds): Harris, K.F. and Maramorosch, K. in Aphids as Virus Vectors, Academic Press, New York, pp. 253–273.
85. Rochow, W.F., 1979, Field variants of barley yellow dwarf virus: detection and fluctuation during twenty years, *Phytopathology* **69**:655–660.

86. Rochow, W.F., 1982, Dependent transmission by aphids of barley yellow dwarf *Luteoviruses* from mixed infections, *Phytopathology* **72**:302–305.
87. Rochow, W.F. and Ball, E.M., 1967, Serological blocking of aphid transmission of barley yellow dwarf virus, *Virology* **33**:359–362.
88. Rochow, W.F. and Carmichael, L.E., 1979. Specificity among barley yellow dwarf viruses in enzyme immunosorbent assays, *Virology* **95**:415–420.
89. Rochow, W.F. and Duffus, J.E., 1981, Luteoviruses and yellows diseases, (ed): Kuzstak, E. in Handbook of Plant Virus Infections and Comparative Diagnosis, Elsevier/North-Holland, Amsterdam, pp. 147–170.
90. Rochow, W.F. and Gill, C.C., 1978, Dependent virus transmission by *Rhopalosiphum padi* from mixed infections of various isolates of barley yellow dwarf virus, *Phytopathology* **68**:451–456.
91. Ruzicska, P., Combos, Z., and Farkas, G.L., 1983, Modification of the fatty acid composition of phospholipids during the hypersensitive reaction in tobacco, *Virology* **128**:60–64.
92. Slykhius, J.T., 1975, Seasonal transmission of wheat spindle streak mosaic virus. *Phytopathology* **65**:1133–1136.
93. Slykhuis, J.T., 1976, Virus and virus-like diseases of cereal crops, *Annu. Rev. Phytopathol.* **14**:189–210.
94. Slykhuis, J.T. and Polak, Z., 1971, Factors affecting manual transmission, purification, and particle lengths of wheat spindle streak mosaic virus, *Phytopathology* **61**:569–574.
95. Smith, H.G., 1963, Control of yellow dwarf virus in cereals, *N.Z.J. Agric. Res.* **6**:229–244.
96. Steponkus, P.L., 1984, Role of the plasma membrane in freezing injury and cold acclimation, *Annu. Rev. Plant Physiol.* **35**:543–584.
97. Suzuki, M. and Nass, H.G., 1988, Fructan in winter wheat, triticale and fall rye cultivars of varying cold hardiness, *Can. J. Bot.* **66**:1723–1728.

5
Artificial Diets for Blood-Feeding Insects: A Review

John R. DeLoach and George Spates

Introduction

Blood feeding has evolved in four orders of insects, but only in the order *Anoplura* do all members, nymphs and adults, feed exclusively on vertebrate blood, whereas in the order *Siphonaptera* only the adult members feed exclusively on blood. All members, including nymphs and adults, of the family *Cimicidae* and subfamily *Triatominae*, in the order *Hemiptera*, feed exclusively on blood. Blood-feeding dependency is more versatile in the order *Diptera* than in the preceding orders. Blood-feeding dipterans may be categorized into four basic groups; group 1, only the larvae feed exclusively on blood, for example, the Congo floor maggot and certain other Calliphorid species whose maggots feed exclusively on nestling birds; group 2, adult male and females feed exclusively on blood and the larvae are indirectly dependent on blood because they are nourished internally by the mother, for example, tsetse flies and sheep keds; group 3, only the adult females feed exclusively on blood, for example, anautogenous mosquitoes, horse flies, and blood-sucking gnats; and group 4, both sexes are obligatory blood-feeders, examples here are stable flies and horn flies.

The classification of insect diets was proposed first by Doughtery (16). Vanderzant described the development of artificial diets and nutritional requirements for insects (60). However, most of her descriptive summary deals with plant-eating insects and is not appropriate for blood-feeding insects. Vanderzant based the need for dietary studies on nutrition, testing

John R. DeLoach, U.S. Department of Agriculture, Agricultural Research Service, Veterinary Toxicology and Entomology Research Laboratory, College Station, Texas 77840, U.S.A.
George Spates, U.S. Department of Agriculture, Agricultural Research Service, Veterinary Toxicology and Entomology Research Laboratory, College Station, Texas 77840, U.S.A.
© 1991 Springer-Verlag New York, Inc. *Advances in Disease Vector Research*, Volume 7.

compounds for physiological effects, maintaining colonies, and the mass production of insects for various purposes. Diets were classified into three general types. Holidic diets are those in which chemical formulae can be used to describe the ingredients that are compounded. With appropriate descriptions, these diets are referred to as defined diets. A second type of diet is one that contains one or more unrefined substances from animals, plants, or microorganisms. The additive is intended often to provide only one nutrient, although it may contribute several classes of nutrients. A third type of diet consists mainly of crude materials intended to imitate the natural food. This latter class of diet is assumed to have all the necessary nutrients but may contain also nutrients that are not used. These diets are usually inexpensive and can be used for mass rearing.

Classification of Blood Dietary Studies

Dietary studies for adult blood-feeding arthropods can be categorized into similar levels of inquiry. Defibrinated blood, heparinized blood, citrated blood, freeze-dried blood, blood mixtures, and fresh, frozen blood are intended to imitate the natural diet of the host-animal blood. Unnatural hosts' blood also belong in this category.

A second category of diets for blood-feeding arthropods includes diets that contain components of blood. Diets of washed erythrocytes and serum, erythrocytes and serum proteins, erythrocyte proteins and serum proteins, serum lipids and erythrocyte lipids, whole blood and serum proteins, and freeze-dried blood and calf serum are examples of the second category of diets. Dietary studies on tsetse flies and stable flies have been conducted primarily at this level of inquiry. Questions such as superiority of one blood over another have been addressed at this level.

A third category of diets is the defined diet or holidic diet and few studies have achieved this level of inquiry. Defined diets are those in which amino acids, lipids, vitamins, and minerals are described by chemical formulae.

Much research has been devoted to dietary requirements of larval stage mosquitoes. The elegant work of Dadd and others on the nutritional requirements of mosquito larvae is indicative of the efforts with mosquito diets (3–6, 42, 45). Not all adult mosquitoes are obligatory blood feeders. *Anopheles albimanus* has been maintained on frozen stored cattle blood (58). *Culex pipiens* can produce eggs without either whole blood or serum proteins in their diet (65). And adult *Aedes aegypti* produce viable eggs when fed a diet of vitamins and essential amino acids (66).

Rhodnius prolixus feeds on various blood fractions and requires ATP as a phagostimulant (48). Langley et al. reported that triatomine bugs have superior reproduction when reared on pig blood (33).

Recent research to elucidate the dietary requirements for tsetse flies and stable flies has reached Level III of inquiry. The remainder of this paper will describe a portion of the inquiry into the nutritional aspects of dietary studies on tsetse flies and stable flies. The work described herein will not deal with the practical aspects of successful in vitro rearing of flies but will be concerned with dietary studies.

Much of the developments in dietary studies with tsetse flies has been driven from the need to mass-rear male flies for the sterile male release strategy of insect control (1). To this end, a number of investigators have developed technology appropriate for mass-rearing tsetse flies (2, 28, 31, 33, 37–39, 49, 61–65). Developmental studies on artificial diets for these blood-feeding insects would not have been possible without membrane-feeding techniques. The first report of tsetse feeding through a membrane was made in 1912 by Rodhain as quoted by Tarshis (57). Many kinds of membrane-feeding techniques were reviewed by Tarshis. Much of the tsetse dietary studies have been conducted with a silicone-type membrane (62) or an agar–parafilm membrane (28). These techniques depend on fresh animal blood, frozen animal blood or freeze-dried blood.

Tsetse flies are unique among muscoid flies in that only the adult life stage feeds on the host externally. Thus the entire life cycle is dependent of the nutritive value of whole blood. Langley and Pimley (31) determined some effects of diet composition on feeding, digestion, and reproduction in *Glossina morsitans morsitans*. Whole blood was separated into plasma and erythrocyte components and a combination of bovine serum albumin (BSA) with porcine erythrocytes met the nutritional requirements of *G. m. morsitans*. Later, BSA was shown to be required for *G. palpalis palpalis* (55).

Approaches for Dietary Studies

In effect, three approaches have been taken in the development of artificial diets for blood-feeding insects. Whole blood can be separated into basic components and recombined with variations subsequently fed to flies, and the reproductive efficiency of flies can be monitored. Thus sequentially eliminating nonessential blood components is one way to define the essential nutrients of blood. This approach was used quite successfully in the 1970s at the Tsetse Research Laboratory, at Bristol in England (28–33). Another approach is defined less clearly and involves a more random combination of dietary components such as essential amino acids, vitamins, and lipids. A third approach is a derivative of the sequential elimination and combination of natural dietary components. This approach involves specific chemical analysis of each dietary component and is more time-consuming and costly, but this approach is more likely to lead to a holidic diet.

Composition of Blood

Analyses of natural diets such as whole blood are numerous in the literature. For example, all known constituents of human blood have been defined (19). Blood comprises red blood cells, white blood cells, platelets, fluid, salts, minerals, intermediary metabolites, hormones, and various proteins. Of the total cell content of blood, greater than 99.9% is contributed by red blood cells. Therefore, white blood cells comprise less than 1% of the cellular component. For their primary functions, they contribute an essential component of life. However, white blood cells are not thought to be an essential dietary component for tsetse flies. Red blood cells contain one major protein, hemoglobin, which comprises over 90% of the total cellular protein. Red blood cells also contain lipids, sugars, nucleosides, intermediary metabolites, and other proteins. Plasma consists of protein, lipids, and sugars. The major protein of plasma is albumin.

Development of Artificial Diet for Glossina and Stomoxys

Stable flies have been laboratory-reared successfully for many years on citrated bovine blood (15). Adult stable flies are obligatory blood-feeders and require three or more blood meals for mating and reproduction (17, 30, 36). Stable flies reportedly have higher fecundity when fed in vitro on blood from herbivores (cattle, sheep, goat, horse) than on blood from omnivores (pigs) (53). And stable flies require components from both serum and red-blood-cell fractions (59). However, due to the ready availability of cow blood from abattoirs, little research was conducted beyond Level I studies for many years.

Many reports in the literature have been concerned with laboratory rearing conditions and, to a lesser extent, nutritional adequacy of the diet of tsetse flies. Generally, such studies were characterized in a report by Langley and Pimley who pointed out that poor reproductive performance of *G. m. morsitans* fed in virto was due mainly to quantitative aspects of feeding rather than qualitative deficiencies in the blood diet (31). For example, citrated bovine blood was reported to be toxic to *G. m. morsitans* (29). Defibrinated bovine blood for in vitro feeding of *G. m. morsitans* was inferior to feeding on cattle (49). Langley reported that defibrination of blood may diminish its nutritive value (28). Later, fresh defibrinated bovine blood was reported to be superior to both hemolyzed or freeze-dried, defibrinated blood (37). Nevertheless, several species of tsetse have been reared successfully in vitro on whole blood. *G. austeni* has been reared on fresh pig blood (38). *G. p. palpalis* and *G. m. morsitans* can be maintained in vitro on fresh cow blood but not on horse blood (64).

In spite of the many successful in vitro rearing techniques for different tsetse species, the question of membrane feeding versus host-animal feeding having qualitative effects on tsetse reproduction, and survival was only answered in 1982 (11). Once the *in vitro* feeding technique was validated at least for *G. p. palpalis* on guinea pig blood, qualitative dietary studies could be conducted without a concern for quantitative aspects of feeding technique.

Level I in vitro studies for tsetse flies have included different animal bloods as reported by Langley et al. (31, 33) and Moloo and Pimley (39). Fecundity and pupal weights were highest on animal bloods in the following order: pigs > sheep > goat > cow > horse. Additives to diets were unsuccessful in duplicating the reproductive response of flies fed on pig blood. Other Level I studies were conducted also with *G. morsitans* by Langley et al. (33) in which erythrocytes were combined with various albumins and in which blood mixtures were used. Level II and Level III inquiry began with the isolation of specific protein components and recombination into suitable diets (20, 21, 54–56).

Bovine serum albumin was shown to be necessary for normal ovarian development in *G. p. palpalis* (54–56). BSA was shown also to be required for *G. m. morsitans*. Moreover, delipidation of BSA impaired reproduction in *G. m. morsitans* (21). Later, lipids associated with BSA were shown to be required by tsetse for reproduction (23). Studies also showed the requirement for red-blood-cell fraction of blood (39).

Efforts to determine the nutritional superiority of porcine blood to bovine blood for both tsetse flies and stable flies have been inconclusive (12, 13, 34). The major differences in the two bloods lie in the erythrocytes. Porcine erythrocytes contain phosphatidylcholine (PC), whereas bovine erythrocytes do not contain PC (40, 41). An amino acid analysis of both bloods reveal significant differences. Bovine blood has higher levels of phenylalanine (8). Moloo and Pimley (39) reported that the addition of high levels of free isoleucine, threonine, or an amino acid mixture were deleterious to *G. morsitans*. The added amino acids used by Moloo and Pimley were 500- to 1500-fold the concentration of free amino acids in freeze-dried blood (8).

DeLoach and Spates (10) attempted to explain the importance of amino acids in their hemoglobin (Hb) plus BSA dietary studies. A 6.5% BSA diet is deficient in the essential amino acids phenylalanine, leucine, methionine, valine, and histidine. Hb (15.4%) solution was deficient in only the amino acid isoleucine. A diet of Hb plus BSA (22.8% total protein) was not deficient in any amino acids when compared with bovine blood. However, amino acid analysis alone is insufficient to explain the requirements for BSA and Hb.

A long-term study of the nutritional superiority of pig blood over cow blood for *S. calcitrans* revealed several interesting features (50). Flies fed for 10 consecutive generations had a higher fecundity on pig blood than on cow blood. Flies fed freeze-dried blood had a consistently higher fecundity than did flies fed fresh blood. The source of freeze-dried blood was unimportant because blood obtained from abattoirs in Austria or Texas and subsequently freeze-dried were equally nutritive for the flies. In fact, flies fed freeze-dried or fresh pig blood set a new benchmark for fecundity of stable flies (50).

Possible correlations between reproductive performance and lipid content of blood diets for *S. calcitrans* have been established (51). No correlation exists between fecundity and total cholesterol or total lipids in diets. The most significant finding from these studies is that bovine blood, which gave low fecundity, had unusually high concentrations of the C18:3 fatty acid. Addition of linoleic acid to bovine blood for feeding to tsetse flies has been reported previously to increase tsetse fly mortality (39). Linoleic acid is an essential fatty acid derived from plants in the diet of cattle. Thus high concentrations in blood are probably a reflection of the animal's diet. Addition of exogenous cholesterol to bovine blood is also known to be detrimental to fecundity of tsetse flies (12). Thus the quality and the quantity of lipids appear to be important to the flies. When varying concentrations of liposome-borne cholesterol were fed to *G. p. palpalis*, a higher mortality was associated with higher concentrations of cholesterol (J.P. Kabayo and J.R. DeLoach, unpublished). A biochemical basis of the nutritional quality of tsetse fly diets appeared elusive. Kabayo et al. analyzed diets for amino acid, triglyceride, and cholesterol content (22). In general, diets with high amino acid content, that is, higher percent of protein, had a nutritionally higher quality. However, there was no correlation between nutritional quality and endogenous triglyceride or cholesterol content.

Consideration for development of artificial diets for tsetse flies must take into account the effective osmotic pressure of the dietary solution. Osmotic pressures of blood range from 280 to 310 mOsmol and that of suitable diets falls into a similar range (8). Osmotic pressures > 450 mOsmol appear to be detrimental to tsetse flies and stable flies (J.P. Kabayo and J.R. DeLoach, unpublished). Thus artificial diets must be constituted in a way that presents acceptable osmotic pressures.

The physical nature of blood was thought originally to be an important factor in dietary requirements for tsetse flies. Langley et al. (33) reported the poor performance of flies fed on hemolyzed blood when compared with nonhemolyzed blood. However, the notion of utilizing reconstituted freeze-dried blood would seem to contradict these studies. In some instances, freeze-dried blood has been equally nutritive as fresh whole blood for tsetse flies (63), but other research data seem to indicate otherwise (12, 13). Efforts to improve the fecundity of tsetse flies fed on reconstituted freeze-dried blood have included additives (13) and sonication of the reconstituted product (7). Some evidence was given that sonication of reconstituted freeze-dried blood improved the physical character of the blood, that is, solubility.

A series of experiments were conducted by Kabayo to determine the necessary serum or plasma components of blood. Basically, albumin is required for nutritional adequacy. We approached our dietary studies with several assumptions. Porcine erythrocytes are nutritionally superior to bovine erythrocytes (12). Albumin is a requirement but is not clearly defined in dietary studies (54–56). Hemoglobin is also a requirement but its importance is not clearly defined in dietary studies either (59). The first reported Hb and BSA diet for *Glossina* and *Stomoxys* was independently developed by two laboratories (10, 25). The Kabayo diet named *KT-80* consisted of commercially available Hb and BSA, whereas, the College Station diet consisted of laboratory-prepared Hb and commercially available BSA. The KT-80 diet also contained phosphate buffer, bicarbonate, $MgSO_4$, $CaCl_2$, KC1, and glucose as well as the phagostimulant ATP (18). Despite the slight differences of minerals, the major differences in the two so-called "artificial diets" was attributed to the Hb preparations.

An analysis of the lipid content of different Hb preparations, both laboratory and commercial, revealed up to 2 mg of lipid per ml of diet (Table 5.1). Fresh whole blood contained about 4 to 5 mg of lipid per ml of blood. A 6.5% (w/v) BSA solution contained 0.8 mg of lipid per ml. Thus both BSA and Hb preparations are contaminated with lipids, which are an important factor in dietetic research and development of a defined diet would require large quantities of lipid-free HB.

Diets of Lipid-Free Hemoglobin

Research has been conducted to isolate ultrapure Hb devoid of lipids (14, 35, 43, 44). However, until recently there was no methodology available for the isolation of large quantities of lipid-free Hb. Developments in our laboratory have led to a relatively simple process for isolation of kg quantities of Hb (9, 46, 47). The process depends upon hypotonic dialysis of erythrocytes to open membrane pores but not to hemolyze the cells' followed by ultrafiltration through 0.1-µ pores. Approximately 100 g of Hb can be processed in 4 to 6 h with 90% yield of Hb from starting material. Hemoglobin prepared by this process is lipid-free (Table 5.2) and protein purity by high-pressure liquid chromatography (HPLC) and gel electrophoresis is greater than 99% (47).

An important facet of the Hb research is the testing of lipid-free Hb in diets. Experiments with the KT-80 diet (lipid-free Hb instead of Serva Hb) indicates a serious failure of the diet to provide nutritional adequacy for *G. p. palpalis* (J.P. Kabayo and J.R. DeLoach, unpublished). Experiments with lipid-free Hb in *S. calcitrans* diets indicate a complete failure of the diet to provide the necessary nutrients for reproduction (26). Also, tests with *G. m. morsitans* indicate a similar failure of the diet (P.A. Langley and J.R. DeLoach, unpublished). Two questions are raised from these data. Is the failure of the diet due to the absence of lipid from Hb or is the failure due to the fact that the Hb was effectively dialyzed? The second question is important in relation to the data of Kabayo et al. (24). Dialysis of freeze-dried blood, KT-80 diet, or other diets renders the diet unsuitable for *G. p. palpalis*. A dialyzable factor called the *yellow factor* restores the nutritive value of the diet when recombined with the dialyzed diet. The dialyzable component has yet to be identified.

Fundamental questions concerning the components of blood that are absolutely required by tsetse flies and by other blood-feeding flies have been answered only recently. The so-called "artificial diets" developed in Seibersdorf, Austria, and College Station, Texas, both contain protein preparations that are not characterized completely. Protein diets would seem to imply that the flies require no exogenous lipids. However, both dietary proteins contain substantial amounts of lipids. A 6.5%-BSA solution contains up to 0.8 mg of lipid per ml of solution. A 12 to 14 g% Hb solution contains up to 2 mg of lipid per ml of solution (Table 5.1). Hemoglobin, whether laboratory-prepared or commercially obtained, contains substantial amounts of lipids (Tables 5.1 and 5.2). Thus an elusive question concerns the importance of lipids in the artificial diets.

The first definitive dietary feeding experiment with lipid-free Hb and the subsequent establishment of a lipid requirement were conducted with *S. calcitrans*. Tables 5.3 and 5.4 summarizes the results of those studies (27). Flies fed on lipid-free Hb and lipid-free BSA produced no eggs and survival was reduced markedly (Table 5.3). When purified erythrocyte membranes or erythrocyte ghosts were recombined with lipid-free Hb and BSA, fecundity was restored to near normal levels (Table 5.4). These results indicate the absolute requirement for an undefined lipid component(s) in the diet of flies.

Subsequent dietary studies were conducted with lipid components known to constitute the membrane of bovine red blood cells and bovine red blood cell ghosts (27). To validate previous research on the general requirement of lipids, we conducted studies with mixtures of lipid-contaminated Hb (as the source of lipid) and lipid-free Hb added to a

TABLE 5.1. Lipid content of *Stomoxys calcitrans diets*.

Diet content	Total lipids[a] mg/ml
Control: fresh, whole blood bovine	4.0
BSA 6.5 g%	0.8
Dialyzed, freeze-dried bovine Hb	1.9
Dialyzed, freeze-dried porcine Hb	1.6
Nondialyzed, freeze-dried bovine Hb	1.8
Dialyzed, freeze-dried bovine Hb + BSA	3.0
Dialyzed, freeze-dried bovine Hb	2.2
Hb, Sigma	2.0
Hb, Sigma + BSA	1.6

[a] Lipids are for Hb concentrations of 12–13 g%.

TABLE 5.2. Lipid content of bovine hemoglobin preparations.

HB sample	Cholesterol	P^b	SM	PS	PE
Cells hemolyzed in distilled water and processed by ultrafiltration	7.3^a	0.51	14.1	1.71	3.05
Cells hemolyzed in distilled water and centrifuged	2.2	0.32	3.26	0.75	1.84
Dialyzed cells plus ultrafiltration	0.0	0.0	0.0	0.0	0.0

[a] Data are the mean and are expressed as mg of lipid/g Hb. (There were no detectable lipids from the TLC plate.)
[b] P = inorganic phosphorous; SM = sphingomyelin; PS = phosphatidyl serine; PE = phosphatidyl ethanolamine.

TABLE 5.3. Fecundity, egg hatch, and survival of *Stomoxys calcitrans* adults fed on lipid-free hemoglobin.

Diet	No. of eggs/female/day	Hatch (%)	Survival (%)
Freeze-dried bovine blood	69.8^a	97.0^a	95.4^a
Lipid/containing hemoglobin			
plus BSA	48.6^b	86.6^b	94.6^a
plus fatty acid free BSA	47.2^b	75.1^b	93.0^a
Lipid-free hemoglobin			
plus BSA	0^c	0^c	63.2^b
plus fatty acid free BSA	0^c	0^c	59.7^b

[a, b, c] Data are X ($n = 21$). Figures in same column followed by the same letter are not significantly different ($p < 0.05$).

TABLE 5.4. Fecundity, egg hatch, and survival of *Stomoxys calcitrans* adults fed on membrane stroma and protein.

Diet	No. of eggs/female/day	Hatch (%)	Survival (%)
Freeze-dried bovine blood	72.6^a	95.3^a	99.7^a
Lipid-contaminated hemoglobin			
plus BSA	44.7^b	93.1^a	95.5^a
Lipid-free hemoglobin plus BSA	2.7^c	0^b	86.4^b
(1 Vol.) membrane	12.9^d	1.4^c	91.7^b
(5 Vol.) membrane	44.5^b	45.8^b	99.6^a
Bovine plasma	27.5^e	90.1^a	81.2^b
Plasma plus (5 Vol.) membrane	33.0^e	94.1^a	84.7^b

[a, b, c, d, e] Data are X ($n = 21$). Figures in same column followed by the same letter are not significantly different ($p < 0.05$).

constant amount of BSA. These data are summarized in Table 5.5. The results confirmed our previous research that lipids were required by the flies for normal reproduction (26).

Finally, studies were conducted with lipid components of the erythrocyte membrane (Table 5.6). Only a combination of cholesterol, a choline source, and phosphatidyl ethanolamine supported egg production and egg

TABLE 5.5. Survival and reproduction in *Stomoxys calcitrans* fed synthetic diets containing different proportions of hemoglobin-bound lipids.[a]

Diet composition % (w/v in distilled water)			Hb-bound lipid impurities[f]	Survival	Number of	Hatch
BHb	LFHb	BSA	(%)	(%)	egg/female/day	(%)
12	0	6.5	100	97.4 ± 2.6^b	54.5 ± 4.7^b	87.0 ± 6.4^b
10	2	6.5	83	96.2 ± 4.0^b	73.5 ± 3.4^c	94.8 ± 0.9^b
7	5	6.5	58	91.4 ± 3.7^b	71.4 ± 3.1^c	87.6 ± 6.5^b
5	7	6.5	42	98.2 ± 2.0^b	71.4 ± 5.3^c	83.9 ± 6.4^b
2	10	6.5	17	91.2 ± 4.7^b	43.2 ± 3.3^d	33.5 ± 4.1^c
0	12	6.5	0	88.9 ± 4.2^b	0.9 ± 0.4^e	2.1 ± 1.5^d

[a, b, c, d, e] Data are X ± SE (*n* = 21). Data in same column followed by the same letter are not significantly different (*p* < 0.05). BHb = bovine hemoglobin; LFHb = bovine lipid-free hemoglobin.
[f] Based on proportion of lipid-free hemoglobin to lipid-contaminated hemoglobin.

TABLE 5.6. Effect of lipid additives on the nutritional value of a lipid-free synthetic diet to *Stomoxys calcitrans*.[a]

Feeding test	Lipid additives (mg/100 ml lipid-free synthetic diet)					Insect survival (%)	Number of eggs/female/day	Eggs hatched (%)
	SM	PE	PS	CH	Choline chloride			
1	0	0	0	0	0	78.1 ± 9.2^b	0.4 ± 0.2^b	0^b
2	40	0	0	0	0	89.2 ± 3.7^b	25.3 ± 3.3^c	1.6 ± 1.6^b
3	0	7	0	0	0	76.3 ± 4.2^b	10.1 ± 1.6^d	0^b
4	40	7	0	25	0	93.2 ± 6.0^c	28.3 ± 3.8^c	82.5 ± 2.5^c
5	0	0	0	25	40	72.1 ± 3.8^b	26.8 ± 4.2^c	0.9 ± 0.9^b
6	40	7	40	25	0	92.1 ± 4.1^c	36.1 ± 5.8^e	73.0 ± 3.0^c

[a, b, c, d, e] Date are X ± SE of six feeding tests. Data in same column followed by the same letter are not significantly different (*p* < 0.05). SM = sphingomyelin from bovine brain; PE = phosphatidyl ethanolamine; PS = phosphatidylserine; CH = cholesterol.

hatch. The amounts of specific lipid additives used in the dietary tests were similar to levels found in lipid-contaminated Hb (46, 47). Sphingomyelin is a minor class of choline-containing lipids in insects that has not been reported in any dipteran. However, Spates et al. have shown recently that *S. calcitrans* utilize 98% of ingested sphingomyelin from a blood meal (52). Similar findings were reported for tsetse fly *G. morsitans*. Choline has been shown to be an essential dietary component for many insects. Also, it is well known that dipterous insects cannot synthesize choline. Therefore, tsetse flies must obtain all the necessary choline from the blood meal, whereas stable flies can obtain it from the larval stage and blood meal. Because bovine erythrocytes contain no PC in their membrane, the choline is supplied by the sphingomyelin, which is the major phospholipid class in bovine red blood cells (40, 41). Cholesterol is believed to be an essential dietary nutrient for *S. calcitrans*. Spates et al. have shown that 28% of ingested cholesterol is utilized in eggs (52). Also, it is well known that insects are unable to synthesize the sterol nucleus required for synthesis of ecdysteroid growth regulators.

The nutrient requirements of stable flies have not been defined completely. Level III dietary studies have been concerned with purified lipid classes and purified proteins. The fatty acids that are esterified to sphingomyelin were not characterized in our studies and their requirement remains to be determined. Predictions of an amino acid, vitamin, lipid, and mineral diet are not forthcoming because these flies cannot tolerate the high osmotic pressure of such a diet. However, the amino acid content of BSA and Hb are known and there is some redundancy of amino acids between these two proteins. Apparently, both proteins are required, because a diet devoid of either is insufficient for normal reproduction of these insects.

References

1. Application of Sterility Principle for Tsetse Fly Suppression (Review of a Panel) Joint FAO/IAEA Division of Atomic Energy in Food and Agriculture. Paris, June, 1971.
2. Bauer, B. and Aigner, H., 1978, in vitro maintenance of *Glossina p. palpalis, Bull. Ent. Res.* **68**:393–400.
3. Dadd, R.H., 1978, Amino acid requirements of the mosquito *Culex Pipiens*: Essentiality of Aspargine, *J. Insect. Physiol* **24**:25–30.
4. Dadd, R.H. and Kleinjan, J.E., 1976, Chemically defined dietary media for larvae of the mosquito, *Culex pipiens*: Effects of colloid texturizers, *J. Med. Entomol.* **13**:285–291.
5. Dadd, R.H. and Kleinjan, J.E., 1977, Dietary nucleotide requirements of the mosquito, *Culex pipiens, J. Insect. Physiol.* **23**:333–341.
6. Dadd, R.H. and Kleinjan, J.E., 1978, An essential nutrient for the mosquito *Culex pipiens* associated with certain animal-derived phospholipids, *Annu. Entomol. Soc. Am.* **71**:794–800.
7. DeLoach, J.R., 1983, Sonication of reconstituted freeze-dried bovine and procine blood: Effect on fecundity and pupal weights of the tsetse fly, *Comp. Biochem. Physiol.* **76A**:47–49.
8. DeLoach, J.R. and Holman, G.M., 1983, An amino acid analysis of freeze-dried bovine and porcine blood used for in vitro rearing of *Glossina palpalis palpalis* (Rob.-Desv.), *Comp. Biochem. Physiol.* **75A**:499–502.
9. DeLoach, J.R., Sheffield, C.L., and Spates, G.E., 1986, A continuous-flow high-yield process for preparation of lipid free hemoglobin, *Anal. Biochem.* **157**:191–198.
10. DeLoach, J.R. and Spates, G.E., 1984, Hemoglobin and albumin diet for adult *Stomoxys calcitrans* and *Glossina palpalis palpalis, Southwestern Entomol.* **9**:28–34.
11. DeLoach, J.R. and Taher, M., 1982, in vitro rearing of *Glossina palpalis palpalis* (Robineau-Desvoidy) (*Diptera: Glossinidae*) on Guinea-Pig Blood, *Bull. Ent. Res.* **72**:663–667.
12. DeLoach, J.R. and Taher, M., 1983, Investigations on development of an artificial diet for in vitro rearing of *Glossina palpalis palpalis* (*Diptera: Glossinidae*), *J. Econ. Etomol.* **76**:1112–1117.

13. DeLoach, J.R. and Taher, M., 1984, An improved freeze-dried blood diet for membrane feeding of *Glossina palpalis palpalis*, *Comp. Biochem. Physiol.* **79A**:383–386.

14. Devenuto, F., Zuck, T.F., Zegna, M.I., and Moores, W.Y., 1977, Characteristics of stroma-free hemoglobin prepared by crystalization, *J. Lab. Clin. Med.* **89**:509–516.

15. Doty, A.E., 1937, Convenient method of rearing the stable fly, *J. Econ. Entomol.* **30**:367–369.

16. Doughtery, E.C., 1959, Introduction of axenix culture of invertebrate metazon: A goal, *Annu. N.Y. Acad. Sci.* **77**:27–54.

17. DuToit, G.D.G., 1975, Reproductive capacity and longevity of stable flies maintained in different kinds of blood, *J. S. Afr. Vet. Ass.* **45**:345–347.

18. Galun, R. and Margalit, J., 1969, Adenine nucleotides as feeding stimulants of the tsetse fly *Glossina ousteric* newst, *Nature*, London **222**:548–584.

19. Handbook for Biological Fluids, 2nd ed. X, CRC Press, p. 551.

20. Kabayo, J.P., 1979, Dietetics of *G. Morsitans*, *Trans. R. Soc. Trop. Med. Hyg.* **74**:277–278.

21. Kabayo, J.P., 1982, The Nature of the nutritional importance of serum albumin to *Glossina morsitans*, *J. Insect Physiol.* **28**:917–923.

22. Kabayo, J.P., DeLoach, J.R., Spates, G.E., Holman, G.M., and Kapatsa, G.M., 1986, Studies on the biochemical basis of the nutritional quality of tsetse fly Diets, *Comp. Biochem. Physiol.* **83A**:133–139.

23. Kabayo, J.P. and Langley, P.A., 1985, The nutritional importance of dietary blood components for reproduction in the tsetse fly, *Glossina morsitans*, *J. Insect Physiol.* **31**:619–624.

24. Kabayo, J.P. and Taher, M., 1986, Effect of dialyzing diet components on the reproductive physiology of *G. p. palpalis*, *J. Insect Physiol.* **32**:543–548.

25. Kabayo, J.P., Taher, M., and Van der Vloedt, A.M., 1985, Development of a synthetic diet for *Glossina* (*Diptera: Glossinidae*), *Bull Ent. Res.* **75**:635–640.

26. Kapatsa, G., Spates, G.E., Sheffield, C.L., and DeLoach, J.R., 1989, Comparison of lipid-free haemoglobin and stroma-contaminated haemoglobin diets for adults of *Stomoxys calcitrans* (L.) (Diptera:Muscidae), *Bull. Ent. Res.* **79**:41–45.

27. Kapatsa, G.M., Spates, G.E., Sheffield, C.L., Kabayo, J.P., and DeLoach, J.R., 1989, The nature of erythrocyte stroma lipid components required for normal reproduction in *Stomoxys calcitrans* (Diptera: Muscidae), *J. Insect Physiol.* **35**:205–208.

28. Langley, P.A. (1972a) Further experiments in rearing tsetse flies in the absence of a living host, *Trans. R. Soc. Trop. Me. Hyg.* **66**:306–310.

29. Langley, P.A., 1972, The role of physical and chemical stimuli in the development of in vitro feeding techniques for tsetse flies. *Glossina* spp. (Dipt., Glossinidae), *Bull. Ent. Res.* **62**:215–228.

30. Langley, P.A., Ogwal, L.M., Felton, T., and Stafford, K., 1988, Lipid digestion in the tsetse fly, *Glossina morsitans*, *J. Insect Physiol.* **33**:981–986.

31. Langley, P.A. and Pimley, R.W., 1975, Quantitative aspects of reproduction and larval nutrition in *Glossina morsitans morsitans* Westw. (Diptera, Glossinidae) fed in vitro, *Bull. Ent. Res.* **65**:129–142.

32. Langley, P.A. and Pimley, R.W., 1978, Rearing of triatomine bugs in absence of a live host and some effects of diet on reproduction in *Rhodnius prolixus* stal

(Hemiptera; Reduviidae). *Bull. Ent. Res.* **68**:243–250.

33. Langley, P.A., Pimley, R.W., Mews, A.R., and Flood, M.E.T., 1978, Effect of diet composition on feeding digestion, and reproduction in *Glossina morsitans*, *J. Insect Physiol.* **24**:233–238.
34. Lie-Injo, L.E., 1976, Simple method for the isolation and purification of hemoglobin components, *J. Chromat.* **117**:53–58.
35. McGregor, W.S. and Dreiss, J.M. (1955), Rearing stable flies in the laboratory, *J. Econ. Entomol.* **48**:327–328.
36. Meola, R.W., Harris, R.L., Meola, S.M., and Oehler, D.D., 1977, Dietary-induced secretion of sex pheromore and development of sexual behavior in the stable fly, *Environ. Entomol.* **6**:895–987.
37. Mews, A.R., Baumgartner, H., Luger, D., and Offori, E., 1976, Colonization of *Glossina morsitans morsitans* Westw (*Diptera, Glossinidae*) in the laboratory using in vitro feeding techniques, *Bull. Ent. Res.* **65**:631–642.
38. Mews, A.R., Langley, R.A., Pimley, R.W., and Flood, M.E.T., 1977, Large-scale rearing of tsetse flies (*Glossina* spp.) in the absence of a living host, *Bull. Ent. Res.* **67**:119–128.
39. Moloo, S.K. and Pimley, R.W., 1978, Nutritional studies in the development of in vitro feeding techniques for *Glossina morsitans*, *J. Insect Physiol.* **24**:491–497.
40. Nelson, G.J., 1967, Composition of neutral lipids from erythrocytes of common mammals, *J. Lipid Res.* **8**:374–379.
41. Nelson, G.J., 1967, Lipid composition of erythrocytes in various mammalian species, *Biochem. Biophys. Acta* **144**:221–232.
42. Prasad, R.S., 1987, Nutrition and reproduction in haemataphagous arthropods, Proc. Indian Acad. Sci. **96**:253–273.
43. Rabiner, S.F., O'Brien, K., Peskin, G.W., and Friedman, L.H., 1970, Further studies with stroma-free hemoglobin solution, *Annals of Surg.* **171**:615–622.
44. Rosenberry, T.L., Chen, J.F., Lee, M.M.L., Moulton, T.A., and Onigman, P., 1981, Large scale isolation of human erythrocyte membranes by high volume molecular filtration, *J. Biochem. Biophys. Methods* **4**:39–48.
45. Sheffield, C.L., Spates, G.E., and DeLoach, J.R., 1987, Preparation of lipid-free human hemoglobin by dialysis and ultrofiltration, *Biotechnol. Appl. Biochem.* **9**:230–238.
46. Sheffield, C.L., Spates, G.E., and DeLoach, J.R., 1989, Hypoosmotic dialysis and ultrafiltration technique for preparation of mammalian hemoglobin: A comparison of three species, *Biomat. Art. Cells, Art. Org.* **16**:887–904.
47. Singhe, K.R.P. and Brown, A.W.A. Nutritional requirements of *Aedes aegypti, L. J. Ins. Physiol.* **1**:199–220.
48. Smith, J.J.B., 1979, The feeding response of *Rhodnius prolixus* to blood fractions and the role of ATP, *J. Exp. Biol.* **78**:225–232.
49. Southton, H.A.W. and Cochings, K.L., 1961, Laboratory maintenance of *Glossina, Rep. E. Afr. Trypan Res. Org.* 1962–1963, p. 31–33.
50. Spates, G.E. and DeLoach, J.R., 1985, Reproductive performance of adult stable flies (*Diptera*: Muscidae) when fed fresh or reconstituted freeze-dried bovine or porcine blood, *J. Econ. Ent.* **78**:856–859.
51. Spates, G.E. and DeLoach, J.R., 1986, Possible correlations between reproductive performance of *Stomoxys calcitrans* and lipid content of blood diets, *Comp. Biochem. Physiol.* **83A**:667–671.

52. Spates, G.E., DeLoach, J.R., and Chen, A.C., 1988, Ingestion, utilization and excretion of blood meal sterols by the stable fly, *Stomoxys calcitrans*, *J. Insect Physiol.* **34**:1055–1061.
53. Sutherland, B., 1978, Nutritional values of different blood diets expressed as reproductive potentials in adult *Stomoxys calcitrans* L. C. (Diptera: Muscidai), *Onderstepourt J. Vet. Res.* **45**:209–212.
54. Takken, W., 1980, Influence of serum albumin on fecundity and weight of the progeny of the tsetse fly *Glossina palpalis palpalis*, *Entomol. Exp. Appl.* **27**:278–286.
55. Takken, W., 1980, The effect of an albumin deficient diet on the reproductive performance of *Glossina palpalis palpalis* females (Diptera; Glossinidae), *Proc. K. Ned Akad Wel Sev C Biol Med Sci.* **83**:387–397.
56. Takken, W., 1981, Dynamics of follicle development and lipid storage in the tsetse fly *Glossina palpalis palpalis* fed on serum albium deficient diets, *J. Insect. Physiol.* **27**:113–119.
57. Tarshis, I.B., 1958, Feeding techniques for bloodsucking arthropods, *Proc. 10th Int. Cong. Entomol.* **3**:767–784.
58. Thomas, J.A., Bailey, D.L., and Dane, D.A., 1985, Maintenance of *Anopheles albimanes* in frozen blood, *J. Am. Mosq. Controls Assoc.* **1**:538–540.
59. Tuttle, E.L., 1961, Studies of the effect of nutrition on survival and oviposite of laboratory reared stable flies, *Stomoxys calcitrans, L. Dissertation Abstrac.* **22**:1334.
60. Vanderzant, E.S., 1974, Development, significance, and application of artificial diets for insects. *Ann. Rev. Ent.* **19**:139–160.
61. Wetzel, H., 1978, Comparison of colony performance and generation sequence in the rearing of *Glossina p. palpalis* (Rob.-Desv.) using in vitro feeding (Diptera: Glossinidae), *Zeitschr. Angewandte Zool.* **65**:278–286.
62. Wetzel, H., 1979, Artificial membrane for in vitro feeding of piercing-sucking arthropods. *Entomol. Exp. Appl.* **25**:117–119.
63. Wetzel, H., 1980, The use of freeze-dried blood in the membrane feeding of tsetse flies *Glossina p. palpalis*, (Diptera: Glossinidae), *Tropendmed. Parasit.* **3**:259–274.
64. Wetzel, H. and Luger, D., 1978, In vitro feeding on the rearing of tsetse flies. (*G. m. morsitans* and *G. p. palpalis*), *Taopenred Parsit.* **29**:239–251.
65. Wetzel, H. and Luger, D., 1978, The use of deep frozen, stored bovine blood for in vitro feeding of tsetse flies, *Z. Parasitenkd.* **57**:163–168.
66. Woke, P.A., 1937, Effects of various blood fractions on egg products of *Aedes aegypti* Linn, *Am. G. Higg.* **25**:372–380.

6
Transmission of African Trypanosomiasis: Interactions Among Tsetse Immune System, Symbionts, and Parasites

Ian Maudlin

Introduction

The threat posed to man and his domestic livestock by trypanosomiasis in Africa is related almost entirely to the distribution of infected tsetse flies. The natural infection rates of tsetse have long been used by those concerned with epidemiology and control to assess the risk or "challenge" created by the disease. Infection rates are determined by dissection and microscopy of testse; such examinations were (and still are) carried out across the fly belts of Africa—a typical example being the work of Pires et al. (102) who dissected more than 18,000 flies. The following infection rates were recorded for the three major pathogenic trypanosome groups: *T. vivax* 6%, *T. congolense* 2%, and no *T. brucei* infections (102). Jordan (60) brought together the results of field dissections such as these carried out in various parts of Africa between 1964 and 1972 and revealed that low levels of infection were the norm for all three major trypanosome groupings but especially for the *brucei* group, "mature" (salivary gland) infections of which were rarely seen. Localized "hot spots" were observed occasionally for *vivax* group infections.

These natural infection rates were thought to be determined largely by the biology of the parasite; having entered the fly in the blood of its mammalian host the probability of the trypanosome completing its life cycle within the fly would depend on the complexities of the changes involved, which in turn related to the species involved. *Trypanosoma vivax* is confined to the fly mouthparts, and having the simplest life cycle would be expected to have a greater chance of success in the fly than would *T. congolense*, which goes through a midgut stage before returning to mature

Ian Maudlin, Tsetse Research Laboratory, ODA/University of Bristol, Langford, Bristol BS18 7DU, U.K.
© 1991 Springer-Verlag New York, Inc. *Advances in Disease Vector Research*, Volume 7.

in the mouthparts. As it has the most complex life cycle within the fly, *T. brucei* should have the smallest chance of successful cyclical transmission (10). As we have seen, this view was supported by field data: *Trypanosoma vivax* infection rates in tsetse were usually greater than *T. congolense*, which, in turn, were greater than *T. brucei* group infections. (*T. vivax* and *T. congolense* are nonpathogenic to man; the *brucei* group includes *T. b. brucei*, which is pathogenic to domestic livestock, and the two human pathogens [causative agents of sleeping sickness] *T. b. rhodesiense* and *T. b. gambiense*, which are morphologically indistinguishable from *T. b. brucei*.) Although it was logical to relate infection rates to behavioral differences between trypanosome species, this did not account for the intriguing fact that infection rates in wild flies, irrespective of the species of trypanosome, were very low. This is the issue I wish to address—why are infection rates so low in natural fly populations? In answering this question, other mysteries of the fly/trypanosome relationship will be explored.

Experimental Studies of Tsetse Infection Rates

Early experiments were designed to demonstrate whether tsetse were indeed the vectors of sleeping sickness (5). Details of trypanosome life cycles within the fly were elucidated then for *T. gambiense* (106), for *T. vivax* and *T. congolense* (72). The involvement of the peritrophic membrane in the trypanosome life cycle was demonstrated by Wigglesworth (123) and Hoare (50), and the details are worth outlining. In tsetse, the peritrophic membrane is secreted by cells of the proventriculus and consists of an annular sheet of chitin with associated protein (124). For *congolense* and *brucei* species, which have a midgut stage, the peritrophic membrane plays a major role—bloodstream-form trypanosomes enter the gut inside the membrane (endoperitrophic space) and then transform into procyclic forms and establish themselves as midgut or "immature" infections—in the ectoperitrophic space between the midgut cells and the bloodmeal. How trypanosomes pass from the endo- to the ectoperitrophic space is debatable. It has been suggested that the parasites enter this space at the anterior end of the gut where the peritrophic membrane is being secreted (34). A more obvious route to the ectoperitrophic space, and the classical explanation, is that trypanosomes pass round the free end of the membrane in the hind gut (129). This route has been said to be impassable as the pH (5.8) of the hindgut was thought to be lethal to trypanosomes (8). This may be the case with the bloodstream-form trypanosomes but these transform within a few hours (112) in the fly into procyclic forms. Procyclic forms can survive *in vitro* at pH 5.5 (I. Maudlin & S.C. Welburn, unpublished) and the classical route via the hindgut seems most plausible.

Trypanosomes subsequently leave this midgut shelter and migrate to the mouthparts (*congolense*) or salivary glands (*brucei*) where they become "mature" (i.e. mammalian infective). The details of the route that trypanosomes take during this return migration are still, after all this time, shrouded in mystery. Trypanosomes are thought to re-enter the anterior part of the midgut at the point where the peritrophic membrane is secreted and said to be fluid (49). This fluidity may be illusory and recent work suggests that such membranes are, even at this stage, impassable (70). It has been postulated also that *brucei* group trypanosomes may reach the salivary glands by crossing the hemocoel (26) but, as we shall see, this now seems unlikely (14).

The details of the trypanosome life cycle in the fly having been more or less established, attention of laboratory work then turned to the factors that could influence this process. Many experiments have been performed for this purpose the details of which have been reviewed extensively (10, 60, 92). Jordan (60) suggested six major factors that influence fly infection rates: (1) temperature; (2) age of fly at the time of infection; (3) age structure of fly population; (4) hosts of flies; (5) species of fly; (6) variation between individual flies. It is relevant to our current understanding of this subject that we consider the evidence that relates to these and other variables.

Temperature

There is evidence to suggest that natural fly infection rates vary with mean annual temperature; Ford and Leggate (33) showed a positive correlation between tsetse-infection rates and the distance from the median of the fly belt in Africa—which relates to increasing temperature. Such temperature effects were considered to be important in sleeping sickness epidemics that showed seasonal periodicity (28). This periodicity was thought to result from increased man–fly contact brought about by the dry season when fly distribution is restricted to riverine vegetation (2). There is laboratory evidence that raising the temperature at which pupae are kept increases infection rates of flies infected on emergence—*T. b. rhodesiense* (9, 30); *T. vivax* (31); *T. b. brucei* (58). This phenomenon has not been explained fully except that it was presumed that flies emerging at higher temperatures would be more eager to take a feed on the day of emergence—the "day-one" factor to which we will now turn in considering the effect of fly age on infection rates.

Fly Age

Early attempts to infect flies in the laboratory indicated that it was important to use young flies or, more significantly, teneral flies (flies that had not yet taken a bloodmeal). Van Hoof et al. (114) found that groups of teneral *G. p. palpalis* infected with *T. b. gambiense* produced salivary gland infection rates as high as 59%, whereas flies that had taken a previous feed did not exceed 3%. More importantly, Duke (24) showed that *G. p. palpalis* infected repeatedly with *T. b. rhodesiense* or *T. b. gambiense* produced almost identical infection rates (4.1%) to flies infected only once as tenerals (4.8%). This phenomenon was investigated further by Wijers (125) who showed that *G. p. palpalis* infected with *T. b. gambiense* within 24 h of emergence had significantly higher salivary gland infection rates (7.6%) than those fed on the second (1.1%) or third days (0.0%) postemergence—not only was it important that flies were teneral but also that they were infected as soon as possible after emergence. Harley (45) obtained similar results with *T. b. rhodesiense* infections in *G. f. fuscipes*—flies less than 1-day-old had much higher salivary gland infection rates (26.2%) than did those 1 to 2 days old (11.9%). Harley (45) also showed that infection rates in nonteneral flies were greatly reduced—a single gland infection was found in 189 flies infected when 20 days old. As Wijers (125) pointed out, only flies taking their first feed from an infected man had a good chance of becoming vectors, which is supported by the very low salivary gland infection rates found in wild flies even during epidemics of the disease. Recently, it has been suggested (36) that older flies could develop much greater salivary gland infection rates than previously thought possible. It was claimed that comparable salivary gland infection rates could be obtained with *T. b. rhodesiense* in teneral *G. m. morsitans* and 21-day-old flies. This was the case when selected groups are compared but a comparison of the mean infection rates shows tenerals with 16% and nonterals with 8% (this difference is not significant but numbers of flies were not great). Gingrich et al. (36) also found tenerals infected more than 24 h old had a higher salivary gland infection rate (23%) than did tenerals infected within 24 h of emergence (12%), although the midgut-infection rate was still higher in the latter group. The experimental technique used to infect flies in these experiments, that is, reducing the serum content of the infective feed, as we shall see later would have increased artificially midgut infection rates in fed flies (37, 83). Makumyaviri et al. (74) also investigated the effect of age on *T. b. brucei*-infection rates in *G. m. morsitans* and found that teneral flies less than 32 h old produced higher infection rates than did flies 32 to 72 h old. By contrast, nonteneral flies starved for 1 to 2 days before infection did not yield a single infection, but nonterals starved for 48 to 72 h were infectable, although rates were low. Gingrich et al. (36) also found that increasing the period of starvation of nonterals could increase infection rates. Gooding (39) has also shown that nonteneral flies can be infected with *T. brucei* but without periods of starvation before infection. Although fly numbers were small in these experiments (39), teneral *G. m. morsitans* had a greater midgut-infection rate (42%) than did nonteneral flies (27%) despite the fact that flies were fed several times each week on infected animals.

Although it was accepted that *brucei* group trypanosomes established themselves more readily in teneral flies, the same rule was thought not to apply to *T. congolense* or *T. vivax*. Jordan (61) reviewed the evidence on *T. vivax* infections and noted that, in contrast to *brucei* infections (24), 100% *T. vivax* infections could be achieved by repeatedly feeding flies on infected hosts, suggesting that tenerality was not critical for this trypanosome to establish in flies. In the case of *T. congolense*, generally it was assumed that tenerality was not as important as in *brucei* group infections (61); field data showed that older flies had higher infection rates than young flies, suggesting that they were being infected throughout their lives (44). Experimental evidence on this subject was limited until recently when it was shown by Distelmans and colleagues (18) that only those *G. p. palpalis* fed within 1 day of emergence would produce *T. congolense* infections. Mwangelwa and colleagues (94) found that teneral *G. m. morsitans* produced approximately twice the *T. congolense*-infection rates of previously fed flies.

We may conclude that both *brucei* and *congolense* group infection rates in tsetse are affected significantly by the age of the fly but the case for *T. vivax* infections remains uncertain. However, it is also clear from experimental work that it is not age per se that is the critical factor but whether the fly is teneral or has taken a previous, uninfected feed.

Age Structure of Fly Populations

As a small proportion of nonteneral flies are still capable of establishing a *brucei* or *congolense* infection, it follows that infection rates in natural populations would be expected to increase with fly age. This appears to be the case: in a classical study Harley (43, 44) examined large numbers of female tsetse for trypanosome infection and aged the females by ovarian dissection (110). The percentage of infected flies of all three species (*G. pallidipes*, *G. f. fuscipes*, and *G. brevipalpis*) increased with female age; over 80% of infections were found in flies more than 40 to 50 days old. More recently, Ryan *et al.* (109) and Tarimo *et al.* (111) have also shown that infection rates of *T. vivax* and *T. congolense* increase with fly age. The data of Harley (44) show a sudden increase in *T. congolense*-infection rates in females more than 40 days old but tend to level off thereafter. Interestingly, this levelling off is seen in other data (109, 111) but if flies were being infected continually at the same rate as tenerals we would expect infection rate to continue to rise in a linear fashion with fly age. These field data tend to reinforce the laboratory conclusions that infection when young is a critical factor. Whether infection of nonteneral flies, even at a low rate, is important to the epidemiology of sleeping sickness in Africa (36) is doubtful when we consider that it takes about 28 days for a fly to develop a salivary gland infection. The average survival time for wild *G. m. morsitans* is only 29 days for females and 15 days for males (101) indicating that it is important for teneral flies to receive an infective feed, not only because they are more likely to develop an infection but also that they are more likely to live long enough for that infection to mature.

Sex of Flies

The question of fly sex and infection rate is complicated by the age distribution of wild flies. Under the ideal conditions of the laboratory, females have been shown to outlive males; for example in one trial, all male *G. m. morsitans* died within 100 days (16), whereas in another test 50% females of the same species were shown to survive for more than 100 days (85). To quantify the longevity of flies in the field is not a simple matter because of sampling bias; it could be that males are less attracted to the trapping system than females. This problem was answered in part by Phelps and Vale (101) who used six different sampling methods to estimate the age structure of a natural population and calculated that female *G. m. morsitans* lived on average more than twice as long as males. The problem of aging males in the field has been exacerbated by the lack of any accurate aging system comparable to the ovarian system for females. Recently, a novel system for aging flies of both sexes based on pteridine fluorescence has been developed (69, 71), which may resolve this issue.

Assuming that females outlive males in the wild, we should expect overall infection rates in females to be greater than those in males, especially in the case of *brucei* infections, which are thought to take 20 to 30 days to reach the salivary glands (30) [*congolense* group infections take about 14 days to mature in *G. m. morsitans* (120)]. Harley (44) consistently found female infection rates with *T. vivax*, *T. congolense* and *T. brucei* to be greater than males in all three species to fly dissected in the wild.

These field data do not indicate whether females are more readily infected per se than males of the same species. Recent laboratory data (86) show no difference between sexes of *G. m. morsitans* in *T. vivax* infection rates. The situation with *T. congolense* and *T. brucei* group infections is complicated by the differentiation of midgut and mature infections. Moloo (86) found no consistent differences between the sexes in *T. congolense* infection rates. Mwangelwa et al. (94) found that although female *G. m. morsitans* developed more midgut infections with *T. congolense*, males produced a greater proportion of mature (hypopharynx) infections; closer examination of these data shows that females actually matured a greater proportion of their midgut infections (77%) than males (54%).

The case for *brucei* group infections is more clearcut; there is sufficient evidence from laboratory studies to suggest that male tsetse produce greater salivary gland infection rates than do females of the same species. Experimental infections with *T. b. rhodesiense* in *G. m. morsitans* (9, 30) found that males consistently developed a greater proportion of salivary gland infections than females. Harley (46) observed the same effect in males of three tsetse species infected with *T. b. rhodesiense*. Moloo (86) found that salivary gland infection rates of *T. b. brucei* in male *G. m. morsitans* were higher than females in four out of five tests experimenting with different diets, but the differences were not always significant. More recently, Makumyaviri et al. (74) have shown that *G. m. morsitans* males mature significantly more midgut infections of *T. b. brucei* than do females, whether infected as tenerals or nontenerals. In further experiments (75) with the *salmon* eye color mutant of *G. m. morsitans* it was found that males matured a much greater proportion of their midgut infections (53%) than did females (16%).

Species of Fly

There is a very large body of data in the literature, both from the field and the laboratory, to show that fly species differ greatly in their ability to transmit trypanosomes of whatever species. When considering field data, one notes that effects of species are confounded by the feeding preferences of different fly species (to which we shall return). Laboratory data, however, show clear specific effects. Duke (22) infected two species of tsetse with different stocks of *brucei* group trypanosomes and concluded that for all stocks *G. morsitans* (presumably, *G. m. centralis*) was a significantly better vector than *G. palpalis* (presumably, *G. f. fuscipes*). More recently, Harley and Wilson (47) compared three species for the transmissibility of *T. congolense* and found infection rates of 11.6% (*G. m. morsitans*), 13.2% (*G. pallidipes*), and a significantly lower rate of 2.9% in *G. f. fuscipes*. Moloo et al. (91) also found that *G. f. fuscipes* was a poor vector of *T. congolense* as were four other species (*G. austeni, G. p. palpalis, G. p. gambiensis*, and *G. tachinoides*) when compared with *G. m. centralis* and *G. brevipalpis*. A similar pattern was observed (91) when the same species of tsetse were infected with *T. b. brucei* except that *G. brevipalpis* while developing midgut infections produced few salivary gland infections, a phenomenon previously observed by Harley (46). Moloo (90) has since compared the vectorial capacity of these same species with different stocks of *T. vivax* and found that the effect of trypanosome stock was greater than the effect of tsetse species. However, in the same study (90) *G. m. centralis* and *G. brevipalpis* again produced consistently high infection rates irrespective of trypanosome stock, whereas the other species produced very low infection rates when infected with East African *T. vivax* stocks. The protocol of this experiment (90) can be criticized as flies were fed throughout their lives on infected hosts so that the numbers of infective feeds given was unknown—however, this does serve to emphasize the point that, despite repeated infections, some fly species are almost completely resistant to infection with certain *T. vivax* stocks.

Generally, it may be concluded from a review of these data that the *morsitans* group of flies (with the exception of *G. austeni*), which are savannah dwellers, are good vectors of all trypanosome species. *Palpalis* group flies, which tend to have a riverine distribution, appear to be poor vectors of all trypanosomes except certain West African stocks of *T. vivax*. The forest species of the *fusca* group (e.g., *G. brevipalpis*) appear to be good vectors of *T. congolense* and *T. vivax* but poor vectors of *brucei* group trypanosomes.

A word of caution must be sounded here about extrapolating laboratory infection rates to field flies. In the early days (22) there were no established colonies of tsetse, and flies for experiment were obtained by collecting pupae in the field. In most experiments carried out recently comparing the vector potential of different species (88, 90), the flies involved are from laboratory colonies and may bear little resemblance to the wild populations from which they originated. Exceptionally, the experiments of Harley and Wilson (46, 47) used wild-collected pupae and their results should reflect accurately the performance of field flies.

Hosts and Feeding Preferences of Flies

An infective feed is an obvious prerequisite for a tsetse fly to become infected. Data on infection rates in wild animals are naturally rather sparse as they are difficult to collect but those available show greater variation between host species. As well as the availability of infected hosts, the feeding preferences of the flies have to be taken into account. The bloodmeals of tsetse can be identified immunologically and these analyses have revealed specific differences between tsetse in feeding preference (118). This may be important epidemiologically; for example Allsopp (1) found that although only 16% of game animals examined in the Lambwe Valley, Kenya were infected, 90% of bushbuck (a favored host of *G. pallidipes*) were infected and concluded that the bushbuck was responsible for maintaining the high prevalence of trypanosomiasis in that area of Kenya. It follows from these data that flies that feed preferentially on Bovidae will be infected more heavily than those feeding on other hosts. This was borne out by Jordan (59) in West Africa and Harley (43) in East Africa who both showed a correlation between fly infection rates and the proportions of Bovid feeds taken by the flies. It might be thought also that if flies take all their feeds from infected Bovidae, then all of the flies will be infected. Jordan (59) found just such a population of *G. m. submorsitans* living close to a cattle trade route in Nigeria—nearly all of the flies dissected were infected with *T. vivax*. However, the correlations noted between infection rates and proportions of Bovid feeds (43, 59) did not extend to *congolense* or *brucei* trypanosomes, and other factors must be involved in these species. In considering the *vivax* data, it is also necessary to account for the effect of trypanosome stock. Moloo et al. (90) have shown that when different species of colonized tsetse are fed on West African *T. vivax* nearly all become equally infected, whereas those fed on

East African stocks of *T. vivax* vary greatly in their susceptibility to infection, in the same way as *T. congolense* infected flies. We may conclude that species of tsetse differ significantly in their ability to transmit *brucei* and *congolense* group trypanosomes and that these specific differences extend to certain stocks of *T. vivax*.

Trypanosomes

It is clear from much of the data cited above that trypanosome species vary in a consistent way in their ability to establish themselves in their invertebrate host. As we have seen, this variation has been related to the complexities of the trypanosome life cycle within the fly (61). Trypanosome stock can influence infection rates with *T. vivax* (90) but another phenomenon has been observed in *brucei* group trypanosomes, which Duke (20) has called *transmissibility*. Certain stocks of *T. b. gambiense* failed to produce salivary gland infections in the normal vector *G. palpalis* (presumably, *G. f. fuscipes*), although midgut infections were established in the flies (20). Duke asked the question "Does the responsibility lie with the trypanosome or with the fly or with both?" and we shall consider this in detail later. However, a complicating issue was raised by Robertson (106) who noted that there were definite periods in the course of an infection when the blood was not infective to flies. This led her to conclude that "the short forms of the trypanosome in the mammal are destined to carry on the cycle in the transmitting host." This observation is supported by the fact that certain old laboratory strains of *brucei* group trypanosomes when subjected to many serial passages through rodents became nontransmissible through flies (126) and lose their pleomorphic character becoming monomorphic, that is, long and slender. Vickerman (115) has shown that the short stumpy forms in the bloodstream are nondividing and suggests that they can only continue their life cycle in the fly, to which they are biochemically preadapted. This preadaptation consists of the development of a mitochondrion in preparation for the switch to the amino-acid-based (proline) metabolism that takes place in the fly. The monomorphic stocks lose their ability to activate their mitochondrion and so are unable to adapt to life in the tsetse gut (116). This elegant biochemical explanation is not supported by the observations of Duke and other workers (10) who found that some stocks of *T. b. gambiense* failed to transmit even when recently isolated from patients (20), whereas stocks of *T. b. rhodesiense* always produced salivary gland infections (23). We must conclude that *T. b. gambiense* represents a special case among *brucei* group trypanosomes. Although some *brucei* stocks on serial passage may become monomorphic and nontransmissible, the nontransmissibility of certain new isolates of *T. b. gambiense* is evidently an unrelated phenomenon that has not been explained adequately.

The numbers of trypanosomes taken in at the infective feed have been considered to be a factor influencing infection rates, but experiments designed to test this, using infected animals, have been confounded by two factors: (1) calculating the numbers of trypanosomes taken in by the flies; and (2) differentiating the proportions of stumpy and slender forms in the blood of the animal. Van Hoof (113) tested both factors and found that although the morphology of the trypanosomes were unimportant the transmission index was higher when the parasites (*T. b. gambiense*) were most abundant in the host animal. Baker and Robertson (3) using *T. b. rhodesiense* and *T. b. brucei* found that neither trypanosome number or morphology affected fly infection rates. Wijers and Willett (126) and Page (97) concluded that the numbers of stumpy forms present in the host blood did influence *brucei* infection rates. With the advent of membrane-feeding techniques for tsetse (85), it became possible to quantify the numbers of trypanosomes ingested by the fly at the infective feed. Using this technique, Dipeolu and Adam (17) found that dilution of the infective dose did not affect infection rates of *T. b. brucei* in *G. m. morsitans*. Otieno et al. (96) found that numbers of trypanosomes did not influence salivary gland infection rates of *T. b. brucei* in *G. m. morsitans* but did affect midgut infection rates. Recently, we tested this in our laboratory (82) by diluting the infective feed so that flies could be presented with a single organism. The results show that infection rates with *T. congolense* in *G. m. morsitans* were unaffected until flies received a calculated dose of less than one organism, that is, until the probability of receiving a single organism was small. It may be concluded that a single trypanosome is sufficient to infect a tsetse fly but whether or not this trypanosome has to be stumpy in form is not clear.

Individual Tsetse and Fly Genotype

The most obvious conclusion from all the foregoing information on fly infection rates is that irrespective of fly age, species, sex, host, or infecting organism all of the flies within an experiment do not become infected. This phenomenon was only too apparent to Dr. Duke, who in the 1920s was employed at the Human Trypanosomiasis Institute in Uganda to investigate the causes of one of the greatest epidemics in human history. At the turn of the century, sleeping sickness caused by *T. b. gambiense* was thought to have caused 200,000 deaths (2/3 of the whole population) in the affected area of Uganda (4). Under these circumstances, Duke was under great pressure to clarify the epidemiology of the disease and was criticized in an official review of his work for his failure to explain the fact that "even in the most favorable conditions, only a comparatively small percentage of tsetse ever becomes infected with trypanosomes" (20). At about the same time, Lloyd and his coworkers (73) in Nigeria dissected nearly 17,000 tsetse and found 347 (2.08%) with mature *T. congolense* infections and 7 with salivary gland infections (*brucei* group). Most interestingly, 6 of these 7 *brucei* infected flies were also infected with *T. congolense*—far higher than expectations if the infections were independently distributed (2.08% of 7 equals 0.15 flies—see 10). Duke (21) noted these data which together with his own experience of mixed infections led him to the conclusion that "only a certain limited number of individuals of any tsetse population are able to act as hosts for trypanosomes." It was the possibility that these "gifted individuals" (21) perhaps inherited their gift that stimulated the present author to start breeding experiments with tsetse. *G. m. morsitans* males with known phenotype (whether or not they could develop a midgut infection of *T. congolense*) were mated with females of unknown phenotype and their progeny were reared and similarly infected. The F_1 data yielded the surprising result that susceptibility was inherited maternally (76). As the tsetse fly feeds and protects its offspring during larval development, the reciprocal differences observed (76) could have represented variation in the maternal environment. Such a maternal effect would have been expected to be transient but long-term breeding experiments showed that the reciprocal differences demonstrated in the F_1 generation persisted for many generations. These tests also confirmed that male parental phenotype did not affect the progeny (78) and susceptibility to *T. congolense* infection was therefore an extrachromosomally inherited character in *G. m. morsitans*. Isofemale lines of susceptible and refractory flies were established and tested with different species and strains of trypanosome which showed that susceptibility applied to all *brucei* as well as *congolense* stocks tested (84). These experiments also showed that the maternally inherited character "susceptibility" applied only to the establishment of midgut infections—maturation of these infections appeared to be a character associated primarily with trypanosome genotype. Although over 80% of susceptible *G. m. morsitans* would develop midgut infections of both *congolense* and *brucei* stocks, there were, as in outbred flies, great differences in the proportions of mature infections of these two trypanosome species (84). We have also since shown that the character "susceptibility" applies only to teneral flies—"susceptible" flies that have taken a bloodmeal prior to infection behave in the same way as outbred flies (I. Maudlin & S.C. Welburn, unpublished). These breeding tests have since been repeated in another laboratory using wild *G. m. morsitans* as parental stock and have confirmed that susceptibility to trypanosome infection is inherited maternally (F.M. Bushrod personal communication).

The breeding tests outlined here showed that fly genotype was not involved in the establishment of midgut infections. Makumyaviri et al. (75) using the sex-linked "salmon" eye mutant in *G. m. morsitans* found that flies carrying this allele were more susceptible to midgut infection with *T. brucei*. However, examination of the data of Distelmans et al. (19) by chi-square test shows that the "salmon" mutation did not increase significantly midgut infection rates of *T. congolense* in either female or male *G. m. morsitans*.

The sex differences in *brucei* infections noted by many workers, however, did suggest that genotype might play a part in flies infected with this trypanosome. The possibility of sex-linkage of salivary gland infection was considered by Makumyaviri et al. (75), but it was decided that the data did not support such a model. The problem of separating midgut from salivary gland infection rates has probably confounded the interpretation of data from *brucei* experiments. We have recently shown (82a) that it is the maturation of established midgut infections into salivary gland infections that is sex-linked. We conclude that the establishment of midgut infections of both *congolense* and *brucei* infections is a maternally inherited character in teneral *G. m. morsitans*, whereas the maturation of *brucei* group infections is probably controlled by a sex-linked recessive allele. Breeding experiments now in progress should clarify this latter point. We now turn to consider how susceptibility could operate in the fly.

Mechanisms of Susceptibility to Trypanosome Infection

While discussing susceptibility to trypanosome infection we have rather ignored the fact that most flies in the wild are refractory to infection. Buxton (10) considered the evidence that teneral flies were infected more easily than previously fed flies and offered two explanations: changes caused by the bloodmeal in (1) gut biochemistry or (2) the peritrophic membrane. Much speculation had gone into the role of the peritrophic membrane in susceptibility. Evidence was provided that trypanosomes left the midgut via the semiliquid region of the membrane at the point of secretion around the proventriculus (29), and Willett (127) suggested that trypanosomes entered the fly by the same route through which they left—differences between young and older flies lying in the rate of growth of the membrane, young flies offering a larger area of semiliquid membrane for trypanosome penetration. Older flies were thought to have a slower rate of membrane growth, a fact confirmed by later experimental work (48). These studies ignored the fact that even among teneral flies infection rates were very low. Harmsen (48) in a thoughtful paper examined the various hypotheses and concluded that the main barrier to midgut establishment was the destruction in the midgut of nontransformed bloodstream-form trypanosomes. In order to avoid this fate, it was postulated (48) that bloodstream forms could hide in the crop of the fly and undergo a period of adjustment to the biochemical milieu of the midgut. Flies less than 24 h old were said to have an incompletely formed peritrophic membrane that could not contain a complete bloodmeal and forced young flies to store blood in the crop for at least 1 h postfeeding to allow the membrane to grow. This 60 min was said to allow the trypanosomes time to transform into a form, intermediate between the bloodstream and procyclic forms, which could survive in the fly midgut. In support of this hypothesis, the activity of glucose-6-phosphate was shown to become more temperature-labile in trypanosomes that had been in the fly crop for more than 1 h. However, Harmsen (48) did admit that this evidence alone was insufficient to support his hypothesis that trypanosome survival was predicated on a period of delay in the crop. It was also admitted (48) that the route taken by trypanosomes entering the gut remained unknown.

Recent studies of selected lines of tsetse with differing levels of susceptibility, however, have led us to a better understanding of the biochemical basis of this innate refractoriness. We shall see that the barrier to establishment lies not with the period of transformation in the gut, or with peritrophic membrane growth, but rather "in the biochemical condition prevailing in the gut of the fly which has not previously swallowed blood" (10). As to the route taken by the trypanosome, we have seen that there is no reason to suppose that circumnavigation of the peritrophic membrane does not take place.

Tsetse Immune System and Transmission

Until recently little was known of the relationship between parasites and the vector immune system. The reasons for this were largely practical as dipteran vectors provide small quantities of material for research. It is not the aim of this chapter to provide a detailed analysis of insect immune systems, but some general points, taken from an excellent recent review by Lackie (68), should be made. Both cellular and humoral immune mechanisms have been identified in insects and there is no doubt that both forms of response are mounted when provoked by, for example, invasion of the hemocoel. Much work has been done with transplant immunology in insects and it has been shown that (1) insect immune systems cannot recognize allografts (from individuals of the same species) and that (2) the hemocytic response varies—distantly related species provoking the greatest response. When invading microorganisms are recognized as foreign by the immune system they can be phagocytosed or encapsulated within a group of hemocytes forming a nodule in the hemocoel. A humoral response involving melanization independent of the hemocyte system may be mounted also against invading bacteria or protozoa. The question of "memory" in the insect immune system seems unresolved but the level of hemocytic response can be increased, albeit temporarily, by wounding.

Most research on interactions between insect vector immune systems and parasites have been concerned with cellular responses in vitro. For example, microfilariae have been encapsulated by mosquito (40); and blackfly (42) hemolymph; bacteria and *T. brucei* provoke both cellular and humoral responses when injected into the tsetse hemocoel (63–65). A most important observation concerning the humoral response was made by Pereira and colleagues (99) who showed that lectins from the hemolymph and gut of *Rhodnius prolixus* could agglutinate epimastigotes (insect mid-gut forms) of *T. cruzi* in vitro. Moreover, crop, midgut, and hemolymph lectins apparently recognized different sites on the epimastigote cell surface, whereas trypomastigotes from infected mouse blood were not agglutinated. Pereira (99) made the most interesting and, as we shall see, perceptive suggestion that these lectins might be involved in the differentiation of the parasite in the insect gut. These experimental results (99) were confirmed with another vector, *Triatoma infestans*, intestinal homogenates of which were shown to stimulate differentiation of *T. cruzi* epimastigotes (54). Lectins are a group of proteins/glycoproteins with a range of remarkable biological properties including the ability to bind specific carbohydrates (38). Lectin receptors have been demonstrated on the surface of both *T. cruzi* epimastigotes (100) and trypomastigotes (67). Work with commercially available lectins also showed that procyclic forms of all species of African trypanosomes tested had lectin-binding sites (57, 93). Bloodstream forms of the African trypanosomes, however, did show specific differences in lectin binding; *T. congolense* bloodstream forms were agglutinated (57), whereas *T. brucei* were not apparently agglutinated by a variety of lectins (104). These results may simply reflect clonal differences between *brucei* group stocks as lectin binding has since been demonstrated in *T. b. rhodesiense* procyclic and bloodstream forms (56).

Croft et al. (14) were the first to demonstrate that tsetse hemolymph contained a component that was active against African trypanosomes. This factor reduced the motility of procyclic *T. b. brucei* in vitro and was specific in effect—it had no effect on *Crithidia* or *Leishmania*. This hemolymph factor was shown to have a similar effect on *T. congolense* procyclics (25). Ibrahim et al. (51) then showed that the active factor in tsetse hemolymph was an agglutinin with distinct sugar specificity—a lectin. Extracts of tsetse midguts were also shown to contain a lectin with different specificity from the hemolymph lectin. Tsetse, like triatomine bugs, produced lectins of distinct specificity and African trypanosomes had lectin-binding sites on their surfaces.

Lectins and Refractoriness

In vitro experiments with lectins did not actually show that trypanosome behavior was influenced by lectins produced by the tsetse in vivo. This situation was altered radically by some simple in vivo experiments with lectin inhibitors. It had been shown that the agglutination of red blood cells by tsetse midgut lectin could be inhibited specifically by D-glucosamine (51) and it was decided to test whether this aminated sugar had any effect in vivo. The results were startling—over 90% of *G. m. morsitans* fed 0.06 *M* D-glucosamine for the first 5 days (including the infective feed) were found on dissection 21 days later to be infected with *T. congolense* (79). These infection rates were significantly greater than in control flies fed 0.06 M D-galactose (which had the normal laboratory infection rates of 50–60%) showing that the effect of this sugar was specific. Moreover, this effect was the same when much lower concentrations of D-glucosamine were used (0.0075 M) showing that this was not simply an ionic strength phenomenon. Results with *T. b. rhodesiense* were similar, midgut infection rates were doubled in flies fed D-glucosamine compared with controls fed D-galactose for 5 days. Further experiments (I. Maudlin and S.C. Welburn, unpublished) have since confirmed that the effect of D-glucosamine is specific—a range of sugars (e.g. D-mannose, D-glucose) had no effect on midgut infection rates when compared with D-glucosamine. The possible effect of charge was also excluded using lysine (an amino acid that has a strong positive charge), which also failed to increase midgut infection rates. The effect of D-glucosamine was the same in other tsetse species tested (*G. p. palpalis* and *G. pallidipes*—I. Maudlin and S.C. Welburn; unpublished), and we concluded that the aminated sugar was inhibiting the action of midgut lectin produced by the fly. It followed from these experiments that the normal role of midgut lectin was to kill trypanosomes (whether *T. congolense* or *T. brucei*) entering the tsetse and so prevent the fly from becoming infected. Further in vivo work showed that there were differences between species and selected lines of tsetse in the rate at which trypanosomes were removed from the midgut. These differences corresponded to the amount of lectin secreted by the midguts of the flies, measured by red blood cell agglutination tests, following a bloodmeal (122). It was noted also that teneral flies had very low titres of midgut lectin suggesting that lectin was secreted in response to the bloodmeal. It had been shown that midgut infection rates of *T. b. rhodesiense* (37) and *T. congolense* (83) could be increased simply by reducing the amount of serum in infecting feeds but the mechanism involved was not clear. It was now apparent that lectin secretion was in response to bloodmeal serum.

How was trypanosome killing taking place? Commercially available lectins were known to agglutinate procyclic trypanosomes in vitro (56) as did midgut homogenates from tsetse (79). Electron microscopy showed that procyclic trypanosomes found in the posterior midguts of flies 72 h postfeeding were being killed by lysis (122), presumably as a consequence of lectin action following adhesion to the cell surface, although lysis may not be the direct effect of lectin activity.

These results went some way toward explaining many of the phenomena associated with the infection of tsetse that have been discussed above. The refractoriness of most flies in a population is simply a reflection of the ability of their immune system to remove invading trypanosomes from their guts by secreting a specific lectin that binds to procyclic trypanosomes resulting in lysis and cell death. Variation in susceptibility between species could be related to inherent differences in midgut lectin titre. The increased susceptibility of teneral flies reflected the fact that lectin is secreted in response to the bloodmeal; the fed fly presents a far more hostile midgut environment to invading trypanosomes than does the teneral gut as it is primed already with lectin from previous feeds.

The increased susceptibility to trypanosome infection of selected isofemale lines of *G. m. morsitans* reflected the very low level of lectin activity of these flies in the teneral condition. However, there were no differences in rates of killing between nonteneral "susceptible" and "refractory" lines of *G. m. morsitans*, and we have since shown (I. Maudlin and S.C. Welburn, unpublished) that differences in infection rate between these flies also disappear after the first feed. Maternally inherited "susceptibility" (78), as we have suggested (122), is a phenomenon confined to the establishment of midgut infections in teneral flies.

Lectins and Maturation

Further experiments with sugar inhibitors showed that the action of lectins in tsetse was not confined to killing unwelcome trypanosomes. Inhibiting the action of midgut lectin was found to have unexpected effects on maturation; feeding lectin inhibitor for longer periods of time reduced the proportions of midgut infections that matured (80). There were, however, significant differences between stocks of both *T. congolense* and *T. brucei*, some of which did not mature a single infection when fed D-glucosamine throughout their lives and others in which maturation was simply halved (80). These results suggested that the midgut lectin was responsible in some way for stimulating maturation but that the degree of response was determined by the trypanosome stock (i.e., genotype). It was concluded that these stock differences represented variation in lectin-binding sites on the surfaces of the procyclic trypanosome (80). Addition and removal of lectin inhibitor showed that, at most, 72 h exposure to lectin was required to stimulate maturation of midgut infections of *T. congolense* (120). The midgut trypanosomes were able to transform as soon as inhibitor was removed—independent of the time they had spent in the midgut—indicating that maturation was not a process predetermined by the trypanosome but dependent on fly interactions.

That lectin production in the fly midgut was a response to ingestion of bloodmeal serum had been suggested by earlier experiments with serum-free feeds. When flies were maintained on a diet of washed red cells in saline, a large proportion of the flies developed *T. congolense* midgut infections, but few of these infections matured (83). These results, although inexplicable at the time, paralleled the effects produced by feeding lectin inhibitors (79) and support the view that midgut lectin is switched on in response to serum and that lectin is essential for maturation. It is relevant here to note that Isola and colleagues (55) found that gut homogenates from unfed triatomines would not initiate maturation of *T. cruzi* in vitro, whereas extracts from fed bugs would. In triatomines as well as tsetse trypanosome, maturation is apparently dependent on some sort of lectin signal given by the insect following feeding.

As well as the midgut lectin, tsetse flies were found to produce another lectin that was present in the hemolymph (14) and had different sugar affinities from the midgut lectin (51). Agglutination studies of hemolymph from *G. m. morsitans*, *G. p. gambiensis*, and *G. tachinoides* detected lectin activity but did not show any human ABO blood group specificity (52). In a recent study, however, we have found *G. m. morsitans* hemolymph to be highly specific in its agglutination of human red blood cells, with B group cells showing much greater agglutination than any other group (120a). This agglutination was found to be inhibited by sugars belonging to the galactose/N-acetylgalactosamine group that are known to inhibit anti-B group lectins (38). We decided to repeat the in vivo midgut lectin inhibition experiments with infected flies (80) but use a sugar which agglutination tests had shown would inhibit only the hemolymph lectin. These experiments showed that inhibition of the hemolymph lectin of infected flies with alpha-D-melibiose significantly reduced the maturation of both *T. b. rhodesiense* and *T. congolense* in *G. m. morsitans*. Moreover, midgut infection rates were unaffected by melibiose treatment showing that this treatment was not interfering with midgut lectin activity. We concluded that trypanosome maturation requires a second independent signal from the hemolymph lectin as well as that from the midgut lectin (120a).

How would trypanosomes living in the ectoperitrophic space of the insect receive such a signal unless they entered the hemocoel? Evans and Ellis (27) had shown by electron microscopy that *T. b. rhodesiense* was able to penetrate the midgut cells of the fly, which suggested that such trypanosomes may take the "short-cut" hemocoelic route to the salivary glands. Whether or not trypanosomes regularly used the transhemocoelic route to reach the salivary glands [which seems unlikely because of the effect (14) of hemolymph lectin on procyclic trypanosomes] these electron micrographs showed that procyclic trypanosomes often came to lie next to the basement membrane of midgut cells. This suggests that some procyclic trypanosomes may be in intimate contact with the hemolymph and its lectin.

Here then were the first in vivo experiments to show that an insect vector's immune system played a definite role in the transmission of a disease (79, 80, 120). The surprising part was not that lectins could act to kill trypanosomes in the fly midgut—this was anticipated by in vitro agglutination tests—but that such a complex system should be involved in maturation in the fly when maturation of certain stocks could be achieved apparently with relative ease in vitro. Maturation of *T. b. rhodesiense* (15) in cultures was accomplished originally with explants of tsetse tissue and has been shown recently to occur in cultures grown on insect cell lines (66). *T. congolense* has also been shown to mature in culture (41) with infective metacyclics being produced first in the presence of a mammalian cell line and then without any supporting cell monolayer (107). The ability of trypanosomes to mature in vitro independently of the insect/lectin system has been considered by Isola et al. (54) who found that differentiation of *T. cruzi* epimastigotes to infective metacyclics could be brought about in culture by two means: (1) by depletion of nutrients in the culture medium that leads to morphogenesis during the stationary phase of growth (11) or (2) by the specific effect of a biologically active factor. This factor, contained in triatomine midgut homogenate, was active during exponential growth ruling out the possibility of exhaustion of nutrients in the medium (53, 55). Ross (107) has shown that nutrient concentration is also important in the maturation of *T. congolense* in vitro but in this case depletion of nutrients stopped differentiation. That the insect/lectin environment is superior to in vitro systems in stimulating differentiation of parasites is shown by differences in the time scale of the two systems. Maturation of *T. congolense* in tsetse flies is a very rapid phenomenon when compared with the length of time taken for the same process in vitro. Procyclic forms of *T. congolense* can take 14 days to produce metacyclic forms in culture, whereas in the tsetse fly a mature infection can be seen in 7 days postinfection (120). Similarly with *T. brucei* infections (15), although cultures with tsetse explants produced mammalian infective forms in 18 days, without the explants in the medium, infective forms arose only sporadically after longer periods (35–51 days). We conclude that although the maturation process of trypanosomes may be induced in culture, the process is conducted more efficiently under the influence of specific lectins produced by their normal vector. This raises the question of what constitutes the "normal" vector of a trypanosome species.

The "Normal" Vector and Lectin Interactions

We have seen that there are great differences between tsetse species in their ability to transmit identical trypanosome stocks. For example, if we look at the data of Harley and Wilson (47), we see significant differences between two species in the proportions of flies infected with *T. congolense*. These interspecific differences appear as a variation in the ability of trypanosomes to establish a midgut infection and we conclude that this reflects variation in the output of midgut lectins. For example, *G. p. palpalis*, which tends to be a poor vector of all trypanosome species when compared with *G. m. morsitans*, should secrete more midgut lectin—this was found to be the case (122).

There are, however, complications when we consider certain trypanosome/fly interactions. As we have seen, *T. congolense* infections are transmitted poorly by *palpalis* group flies with *G. p. gambiensis* as an extraordinarily poor vector. Although 10% of this fly species established midgut infections, only 2.6% of these infections matured—the comparable figures for *G. m. morsitans* were 56% midgut infections, 75% of which matured (88). It cannot be the level of midgut lectin output that is preventing the maturation of *T. congolense* in *palpalis* group flies and we suspect that it is interaction between the hemolymph lectin of these flies and the trypanosome stock that is the controlling factor. This may be also the case when we consider the transmission of *T. b. gambiense*. As we have seen, even fresh isolates of this trypanosome are sometimes difficult to transmit in the laboratory although midgut infections can be established. Our experience with this trypanosome is that it is difficult to produce mature infections of any "classical" *T. b. gambiense*—the major genetic grouping characterized by the possession of the isoenzyme marker ALAT 1 and low-rodent infectivity (35). We have failed to obtain salivary gland infections with five "classical" stocks of *T. b. gambiense* in *G. pallidipes*, *G. p. palpalis*, or *G. m. morsitans* despite increasing midgut infection rates by feeding D-glucosamine to promote midgut establishment (I. Maudlin and S.C. Welburn, unpublished). These experiments were conducted with bloodstream-form trypanosomes but it has been shown recently that mature infections of this trypanosome can be obtained if the life cycle is short circuited by using procyclic culture forms to infect the flies (105) (the authors' claim that they have achieved cyclical transmission in these experiments is negated by the fact that they started with the midgut and not the bloodstream form of the trypanosome). As Richner and colleagues (105) only obtained mature infections in *G. p. gambiensis* (failing with *G. m. morsitans* and *G. m. centralis*), they concluded that there was a fly species-specific factor responsible for the maturation of this trypanosome. However, we have been able to obtain mature infections of recent human isolates of *T. b. gambiense* in *G. m. morsitans* starting with procyclic forms (I. Maudlin and S.C. Welburn, unpublished). We conclude that in the case of *T. b. gambiense*, it is not the fly species that is the limiting factor but rather the trypanosome genotype and its interactions first with the midgut lectin that prevents establishment and second with the hemolymph lectin that promotes maturation.

We proposed a model to explain these fly–trypanosome interactions based on lectin-binding sites of the trypanosome surface coat (81); this model now needs modification in view of the involvement of both midgut and hemolymph lectins in maturation. The lectin model we have proposed may explain also the differences between sexes in maturation of *brucei* group infections. The single sex-linked recessive gene hypothesized for the maturation of *brucei* infections may be responsible for the output of hemolymph lectin in some manner as yet unknown.

Tsetse Symbionts

We have discussed the breeding experiments that showed "susceptibility" to trypanosome infection was an extrachromosomally inherited character in *G. m. morsitans* (78, 84). The search was then on for cytoplasmic inclusions that could be responsible for this mode of inheritance. Electron microscopy revealed that symbionts identified as rickettsia-like organisms (RLO) could be detected in the developing eggs of more susceptible females than in those of flies from a refractory line—an association was suggested between RLO and susceptibility to trypanosome infection, although the nature of the relationship was not clear and was not claimed to be causal (77). What was clear from infection experiments was that "susceptibility" was not and all or nothing phenomenon; selected families showed great variation in their susceptibility or refractoriness measured by midgut infection rates (84). Whatever cytoplasmic effect was involved, it had to be quantitative in character. As RLO are transmitted transovarially (98), it would be expected that RLO-infected females would pass different numbers of the bacterium to their offspring to produce this within family variation. It has been shown in rickettsia-infected ticks that females with massive infections infect all of their progeny, whereas those with mild rickettsial infections produced lower percentages of infected offspring (6). The simplest approach to establishing a causal relationship between RLO and susceptibility would have been to remove the bacteria from females and test the susceptibility of their offspring. This has been attempted (I. Maudlin, unpublished) but unfortunately the effect of high doses of antibiotics on tsetse flies is to sterilize females (62). Demonstrating quantitative RLO differences between flies proved technically difficult using conventional staining techniques and an alternative method was needed. As a result of developing a system for culturing RLO taken from tsetse hemolymph (121), a pure sample of these intracellular organisms was obtained and a specific DNA probe produced (119). This probe has been used recently to compare RLO numbers in two colonies derived from the same isofemale line of *G. m. morsitans*, the pupae of which have been reared at different temperatures for 2 years (adult flies of both colonies were maintained at 25°C). The results have shown that teneral midguts from pupae held at the normal temperature of 25°C have many times the RLO load of those from pupae held at 22°C—although RLO could be detected in all flies tested. Infection rates have diverged also in these two colonies, those at 22°C giving mean midgut rates of significantly less (29%) than the 25°C colony (67%) (I. Maudlin & S.C. Welburn, unpublished). These results have shown clearly that reduction in RLO number results in reduced susceptibility to trypanosome infection, and we conclude that RLO does is related to susceptibility. Moloo and Shaw (89) have recently detected RLO by electron microscopy in all individuals of a colony of *G. m. centralis* whether trypanosome infected or uninfected and concluded that RLO are not associated with susceptibility. In view of our work with *G. m. morsitans* and other species, this result is not surprising as we have detected RLO in most individuals of all tsetse colonies we have examined with a DNA probe (119). The conclusion drawn (89) is, however, erroneous; the relationship between RLO and susceptibility is essentially quantitative and cannot be tested by this means. The need is for breeding tests, similar to those done with *G. m. morsitans* (84), to be carried out with other species to detect maternal effects. Having found more RLO in older flies, these authors apparently expected more mature infections in them (89). That older flies have more RLO is to be expected as the bacteria naturally divide throughout the life of the fly—for example, Yen (128) has shown in RLO-infected mosquitoes that the microorganisms divide throughout the life cycle with the most rapid divisions taking place in the larvae. However, the RLO load of old flies is irrelevant to maternally inherited susceptibility, which, as we have emphasized in this text, applies only to the increase in midgut infection rates in teneral susceptible flies. Maturation is an entirely separate phenomenon associated with lectin–trypanosome interactions following the establishment of the midgut infection and is independent of RLO infection.

The problem remains of the means by which RLO could affect susceptibility to trypanosome infection. The peritrophic membrane of tsetse flies is a chitinous membrane (123), which is a polymer made up largely of N-acetyl-D-glucosamine. However, every sixth or seventh residue of this polymer is a D-glucosamine residue (108). Cultures of RLO have been found to produce the enzyme chitinase, which can hydrolyze chitin to release glucosamine; chitinase can be detected also in tsetse midguts (S.C. Welburn, unpublished) and is a constituent of molting fluid (12). We have suggested that the action of chitinase in the fly would be to release D-glucosamine, which would accumulate in the gut of the preemergent fly and inhibit the action of the midgut lectin at the first bloodmeal (81). Flies having most RLO in the larval/pupal stages would release greater amounts of D-glucosamine, becoming "susceptibles."

Strictly matroclinal infections such as RLO must impart some selective advantage to the host if they are to persist in natural populations (32). Isofemale lines of "susceptible" *G. m. morsitans* have a significantly better rate of emergence from pupae than do "refractory" lines (I. Maudlin, unpublished), suggesting that RLO confer some advantage during development. It is the case that RLO were detected by DNA probing in nearly all outbred *G. m. morsitans* from the Langford colony (119); this contrasts sharply with the original population in Zimbabwe (from which the colony was established over 20 years ago), which was found to have only about 30% RLO infected (I. Maudlin, unpublished). A colony of *G. austeni* maintained in our laboratory has been shown by DNA probing to have 17% RLO infected (119) and produces midgut *T. congolense* infection rates of about 10% (I. Maudlin and S.C. Welburn, unpublished). All the natural tsetse populations we have examined have been found to have individuals carrying RLO but the frequency varies between species. It appears that RLO are being maintained in wild and laboratory populations of tsetse at varying frequencies by differing selective pressures and are presumably conferring a significant advantage to their hosts. The fact that all flies in a wild population do not possess RLO (in contrast to some tsetse colonies) suggests that RLO carry a balancing selective disadvantage—susceptibility to trypanosome infection might be a disadvantage to a wild fly as midgut trypanosomes use proline as an energy source (116) as do the flies (7). Laboratory tests of infected flies have produced contradictory results—some work has indicated that infection does not affect the longevity of the fly (87), whereas others have shown that longevity is affected significantly by trypanosome infection (95). It is perhaps not possible to answer these questions of fitness using laboratory flies that are under no physiological stress and we await further field data.

Conclusion

We have seen that natural populations of tsetse flies are well equipped to deal with invading parasites using their nonspecific humoral immune system. The parasites are precluded normally from establishing themselves in the fly midgut unless the flies' immune system is compromised in some way. This may be because the system has not been switched on yet (teneral condition) and the fly has a serious bacterial infection (RLO). The

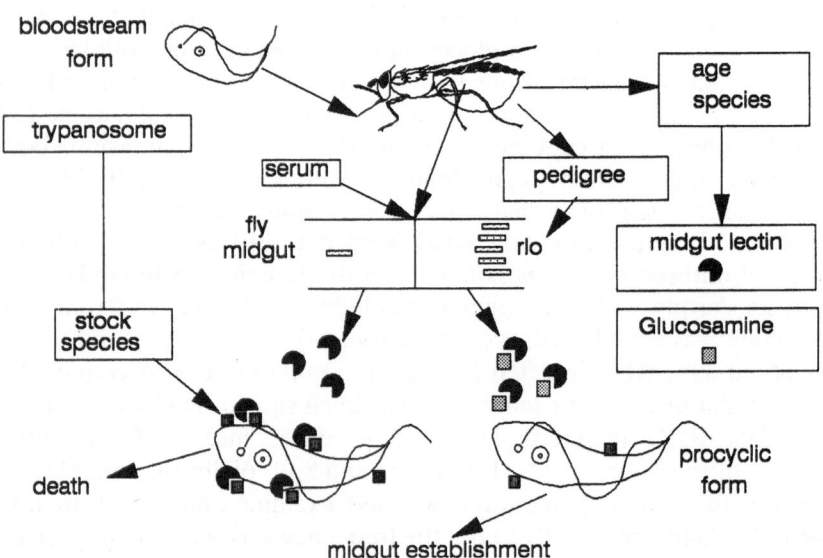

FIGURE 6.1. Model of interaction between fly symbionts (rlo), midgut lectin, and trypanosome stock in the establishment of midgut infections in tsetse.

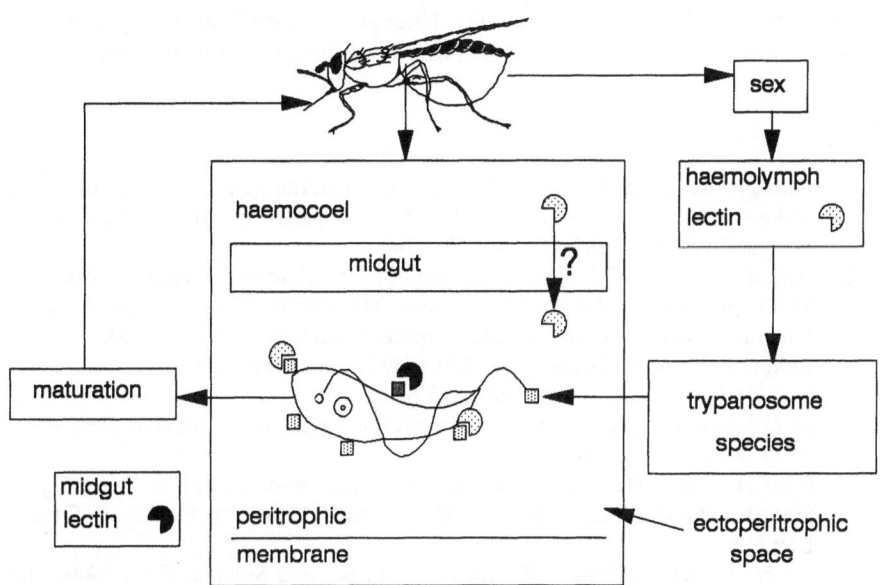

FIGURE 6.2. Model of role of tsetse lectins in the maturation of established midgut infections.

evolutionary response of trypanosomes has been to use these immune molecules to their own ends as stimulants for the maturation process. The model that we propose for trypanosome establishment and maturation in the fly is illustrated in Figures 6.1 and 6.2.

The insect immune system has been implicated in other vector-borne diseases. Lectins from sandfly gut extracts have been shown to agglutinate *Leishmania* spp. (117). Collins et al. (13) have suggested that refractoriness to malaria infection in selected lines of mosquitoes is mediated through encapsulation of the ookinetes in the hemolymph. Work on other vector immune systems, especially if in vivo effects of lectins on transmission can be demonstrated, could be especially rewarding. Given a supply of synthesized lectin, its effects on maturation of in vitro-cultured parasites could be examined. Further knowledge of the biochemistry of tsetse and other insect lectins and eventually the genetic engineering of these molecules may provide the means to improve disease control, for example by blocking transmission through the vector (103).

Acknowledgments. The financial of the Overseas Development Administration of the U.K. Government, through the Livestock Protection

Programme of the Overseas Development Natural Resources Institute, the UNDP/World Bank/WHO Special Programme for Research and Training in Tropical Diseases (TDR), the Joint FAO/IAEA Division of Isotope and Radiation Application of Atomic Energy in Food and Agricultural Development, and The Wellcome Trust is gratefully acknowledged.

References

1. Allsopp, R., 1972, The role of game animals in the maintenance of endemic and enzootic trypanosomiases in the Lambwe Valley, South Nyanza District, Kenya. *Bull. Wld Hlth Org*. **47**:735–746.
2. Apted, F.I.C., 1970, The epidemiology of Rhodesian sleeping sickness, Mulligan, H.W. (ed): in The African Trypanosomiases George Allen & Unwin/Ministry of Overseas Development, London, pp. 645–660.
3. Baker, J.R. and Robertson, D.H.H., 1957, An experiment on the infectivity to *Glossina morsitans* of a strain of *Trypanosoma rhodesiense* and of a strain of *T. brucei* with some observations on the longevity of infected flies, *Ann. Trop. Med. Parasit*, **51**:121–134.
4. Bell, H., 1906, Despatch from Commissioner Hesketh Bell to Secretary of State for the Colonies. *Col. Off. Miscell*. **No. 178**. H. M. Stationery Office, London.
5. Bruce, D., Hamerton, A.E., Bateman, H.R., and Mackie, F.P., 1909, The development of *Trypanosoma gambiense* in *Glossina palpalis*, *Proc. Roy. Soc. (B)*, **81**:405–414.
6. Burgdorfer, W. and Brinton, L.P., 1975, Mechanisms of transovarial infection of spotted fever rickettsiae in ticks, *Ann. N.Y. Acad. Sci*, **266**:61–72.
7. Bursell, E., 1970, Feeding, Digestion and Excretion, Mulligan, H.W. (ed): in The African Trypanosomiases George Allen & Unwin/Ministry of Overseas Development, London, pp. 305–316.
8. Bursell, E. and Berridge, M.J., 1962, The penetration of the ectoperitrophic space of tsetse flies in relation to some properties of the rectal fluid, *Proc. Soc. Exper. Biol*. Jan Meeting. Cited by Freeman in Ref. 34.
9. Burtt, E., 1946, The sex ratio of infected flies found in transmission-experiments with *Glossina morsitans* and *Trypanosoma rhodesiense*, *Ann. Trop. Med. Parasit*. **40**:74–79.
10. Buxton, P.A., 1955, The Natural History of Tsetse Flies, London, H.K. Lewis & Co. Ltd., 816 p.
11. Castellani, O., Ribeiro, L.V., and Fernandez, J.F., 1967. Differentiation of *Trypanosoma cruzi* in culture, *J. Parasitol*. **14**:447–450.
12. Chapman, R.F., 1969, The Insects Structure and Function, London, The English Universities Press, 819 p.
13. Collins, F.H., Sakai, R.K., Vernick, K.D., Paskewitz, S., Seeley, D.C., Miller, L.H., Collins, W.E., Campbell, C.C., and Gwadz, R.W. 1986, Genetic selection of a *Plasmodium*-refractory strain of the malaria vector *Anopheles gambiae*, *Science* **234**:607–610.
14. Croft, S.L., East, J.S., and Molyneux, D.H., 1982, Anti-trypanosomal factor in the haemolymph of *Glossina*, *Acta Trop. (Basel)* **39**:293–302.

15. Cunningham, I., Honigberg, B.M., and Taylor, A.M., 1981, Infectivity of monomorphic and pleomorphic *Trypanosoma brucei* stocks cultivated at 28°C with various tsetse fly tissues, *J. Parasitol.* **67**:391–397.
16. Curtis, C.F. and Langley, P.A., 1972, Use of nitrogen and chilling in the production of radiation-induced sterility in the tsetse fly *Glossina Morsitans*, *Ent. Exp. & Appl.* **15**:360–376.
17. Dipeolu, O.O. and Adam, K.M.G., 1974, On the use of membrane feeding to study the development of *Trypanosoma brucei* in *Glossina, Acta Trop. (Basel)*. **31**:185–200.
18. Distelmans, W., D'Haeseleer, F., Kaufman, L., and Rousseuw, P., 1982, The susceptibility of *Glossina palpalis palpalis* at different ages to infection with *Trypanosoma congolense, Ann. Soc. Belge Med. Trop.* **62**:41–47.
19. Distelmans, W., Makumyariri, A.M., D'Haeseleer, F.D., Claes, Y., LeRay, D., and Gooding, R.H., 1985, Influence of the salmon mutant of *Glossina morsitans morsitans* on the susceptibility to infection with *Trypanosoma congolense, Acta Trop. (Basel)*. **42**:143–148.
20. Duke, H.L., 1930, On the occurrence in man of strains of *T. gambiense* non-transmissible cyclically by *G. palpalis, Parasitology* **22**:490–504.
21. Duke, H.L., 1933, Studies on the factors that may influence the transmission of the polymorphic trypanosomes by tsetse. I. A review of existing knowledge on this subject, with some general observations, *Ann. trop. Med. Parasit.* **27**:99–118.
22. Duke, H.L., 1933, Studies on the factors that may influence the transmission of the polymorphic trypanosomes by tsetse. III. *Glossina morsitans* versus *Glossina palpalis* as a transmitter of the polymorphic group of trypanosomes, *Ann. trop. Med. Parasit.* **27**:123–130.
23. Duke, H.L., 1933, Studies on the factors that may influence the transmission of the polymorphic trypanosomes by tsetse. VII. *T. rhodesiense* versus *T. gambiense*: A comparison of their power to develop cyclically in *Glossina, Ann. trop. Med. Parasit.* **27**:569–584.
24. Duke, H.L., 1935, Studies on the factors that may influence the transmission of the polymorphic trypanosomes by tsetse. IX. On the infectivity to *Glossina* of the trypanosome in the blood of the mammal, *Ann. trop. Med. Parasit.* **29**:131–143.
25. East, J., Molyneux, D.H., Maudlin, I., and Dukes, P. 1983, Effect of *Glossina* haemolymph on salivarian trypanosomes *in vitro, Ann. trop. Med. Parasit.* **77**:97–99.
26. Ellis, D.S. and Evans, D.A., 1977, Passage of *Trypanosoma brucei rhodesiense* through the peritrophic membrane of *Glossina morsitans morsitans, Nature* **267**:834–835.
27. Evans, D.A. and Ellis, D.S., 1975, Penetration of mid-gut cells of *Glossina morsitans morsitans* by *Trypanosoma brucei rhodesiense, Nature* **258**:231–233.
28. Fairburn, H., 1948, Sleeping sickness in Tanganyika territory, 1922–1946, *Trop. Dis. Bull.* **45**:1–17.
29. Fairburn, H., 1958, The penetration of *Trypanosoma rhodesiense* through the peritrophic membrane of *Glossina palpalis, Ann. Trop. Med. Parasit.* **52**:18–19.

30. Fairbairn, H. and Culwick, A.T., 1950, The transmission of the polymorphic trypanosomes, *Acta. Trop. (Basel)*. **7**:19–47.

31. Fairburn, H. and Watson, H.J.C., 1955, The transmission of *Trypanosoma vivax* by *Glossina palpalis*. *Ann. Trop. Med. Parasit*. **49**:250–259.

32. Fine, P.E.M., 1975, Vectors and vertical transmission an epidemiologic perspective, *Ann. N.Y. Acad. Sci.* **226**:173–194.

33. Ford, J. and Leggate, B.M., 1961, The geographical and climatic distribution of trypanosome infection rates in *G. morsitans* group of tsetse flies (*Glossina* Wied., Diptera), *Trans. R. Soc. Trop. Med. Hyg*. **55**:383–397.

34. Freeman, J.C., 1973, The penetration of the peritrophic membrane of the tsetse flies by trypanosomes, *Acta Trop. (Basel)* **30**:347–354.

35. Gibson, W.C., Marshall, T.F. de C., and Godfrey, D.G., 1980, Numerical analysis of enzyme polymorphism: A new approach to the epidemiology and taxonomy of trypanosomes of the subgenus *Trypanozoon*, *Adv. Parasitol.* **18**:175–246.

36. Gingrich, J.B., Ward, R.A., Macken, L.M., and Esser, K.M., 1982, African sleeping sickness: new evidence that mature tsetse flies (*Glossina morsitans*) can become potent vectors, *Trans. R. Soc. trop. Med. Hyg*. **76**:479–481.

37. Gingrich, J.B., Ward, R.A., Macken, L.M., and Schoenbechler, M.J., 1982, *Trypanosoma brucei rhodesiense* (Trypanosomatidae): factors influencing infection rates of a recent human isolate in the tsetse *Glossina morsitans* (Diptera: Glossinidae), *J. Med. Entomol.* **19**:268–274.

38. Goldstein, I.J. and Poretz, R.D., 1986, Isolation, physicochemical characterization and carbohydrate-binding specificity of lectins, Liener, I.E. Sharon, N. and Goldstein, I.J. (ed): in The Lectins Academic Press, Orlando, pp. 35–247.

39. Gooding, R.H., 1988, Infection of post-teneral tsetse flies (*Glossina morsitans morsitans* and *Glossina morsitans centralis*) with *Trypanosoma brucei brucei*, *Can. J. Zool.* **66**:1289–1292.

40. Gotz, P., 1986, Mechanisms of encapsulation in dipteran hosts. Lackie, A.M. (ed): in Immune mechanisms in invertebrate vectors, *Symp. Zool. Soc. Lond.* pp. 1–20. Oxford University Press.

41. Gray, M.A., Ross, C.A., Taylor, A.M. Tetley, L., and Luckins, A.G., 1985, In vitro cultivation of *Trypanosoma congolense*: the production of infective forms from metacyclic trypanosomes cultured on bovine aorta endothelial monolayers, *Acta trop. (Basel)* **42**:99–110.

42. Ham, P.J., Zulu, M.B., and Zahedi, M 1988, In vitro haemagglutination and attenuation of microfilarial motility by haemolymph from individual blackflies (*Simulium ornatum*) infected with *Onchocerca lienalis*, *Med. Vet. Entomol.* **2**:7–18.

43. Harley, J.M.B., 1966, Studies on age and trypanosome infection rate in females of *Glossina pallidipes* Aust, *G. palpalis fuscipes* Newst and *G. brevipalpis* Newst in Uganda, *Bull. Ent. Res.* **57**:23–37.

44. Harley, J.M.B., 1967, Further studies on age and trypanosome infection rates in *Glossina pallidipes* Aust, *G. palpalis fuscipes* Newst and *G. brevipalpis* Newst in Uganda, *Bull. Ent. Res.* **57**:459–477.

45. Harley, J.M.B., 1971, The influence of the age of the fly at the time of the infecting feed on infection of *Glossina fuscipes* with *Trypanosoma rhodesiense*, *Ann. Trop. Med. Parasit.* **65**:191–196.

46. Harley, J.M.B., 1971, Comparison of the susceptibility to infection with *Trypanosoma rhodesiense* of *Glossina pallidipes*, *G. morsitans*, G. fuscipes and *G. brevipalpis*, *Ann. Trop. Med. Parasit.* **65**:185–189.

47. Harley, J.M.B. and Wilson, A.J., 1968, Comparison between *Glossina morsitans*, *G. pallidipes* and *G. fuscipes* as vectors of trypanosomes of the *Trypanosoma congolense* group, *Ann. Trop. Med. Parasit.* **62**:178–187.

48. Harmsen, R., 1973. The nature of the establishment barrier for *Trypanosoma brucei* in the gut of *Glossina pallidipes*, *Trans. R. Soc. Trop. Med. Hyg.* **67**:364–373.

49. Hoare, C.A., 1931, The peritrophic membrane of *Glossina* and its bearing on the life cycle of *Trypanosoma grayi*, *Trans. R. Soc. Trop. Med. Hyg.* **25**:57–64.

50. Hoare, C.A., 1972. *The trypanosomes of mammals* Oxford, Blackwell, pp. 35–47.

51. Ibrahim, E.A.R., Ingram, G.A., and Molyneux, D.H., 1984, Haemagglutinins and parasite agglutinins in haemolymph and gut of *Glossina*, *Tropenmed. Parasit.* **35**:151–156.

52. Ingram, G.A. and Molyneux, D.H., 1988, Sugar specificities of anti-human ABO (H) blood group erythrocyte agglutinins (lectins) and haemolytic activity in the haemolymph and gut extracts of three *Glossina* species, *Insect Biochem.* **18**:269–279.

53. Isola, de E.L., Lammel, E.M., and Gonzalez Cappa, S.M., 1983, Influencia de la hemolinfa de *Triatoma infestans* en la morfogenesis del *Trypanosoma cruzi*, *Rev. Argentina de Microbiologia.* **15**:181–185.

54. Isola, de E.L., Lammel, E.M., and Gonzalez Cappa, S.M., 1986, *Trypanosoma cruzi*: Differentiation after interaction of epimastigotes and *Triatoma infestans* intestinal homogenate, *Exptl. Parasitol.* **62**:329–335.

55. Isola, de E.L., Lammel, E.M., Katzin, V.J., and Gonzalez Cappa, S.M., 1981, Influence of organ extracts of *Triatoma infestans* on differentiation of *Trypanosoma cruzi*, *J. Parasitol.* **67**:53–58.

56. Jackson, P.R. and Diggs, C.L., 1983, *Trypanosoma rhodesiense* bloodstream trypomastigotes and culture procyclic cell surface carbohydrates, *J. Protozool.* **30**:662–668.

57. Jackson, P.R., Honigberg, B.M., and Holt, S.C., 1978, Lectin analysis of *Trypanosoma congolense* bloodstream trypomastigote and culture procyclic surface saccharides by agglutination and electron microscopic technics, *J. Protozool.* **25**:471–481.

58. Jenni, L., 1977, Comparisons of antigenic types of *Trypanosoma (T.) brucei* strains transmitted by *Glossina m. morsitans*, *Acta Trop. (Basel)* **34**:35–41.

59. Jordan, A.M., 1965, The hosts of *Glossina* as the main factor affecting trypanosome infection rates of tsetse flies in Nigeria, *Trans. R. Soc. Trop. Med. Hyg.* **59**:423–431.

60. Jordan, A.M., 1974, Recent developments in the ecology and methods of control of tsetse flies (*Glossina* spp.) (Dipt., Glossinidae)—A review, *Bull. Ent. Res.* **63**:361–399.

61. Jordan, A.M., 1976, Tsetse flies as vectors of trypanosomes, *Veterinary Parasitology.* **2**:143–152.

62. Jordan, A.M. and Trewern, M.A., 1976, Sulphaquinoxaline in host diet as

the cause of reproductive abnormalities in the tsetse fly (*Glossina* spp), *Ent. Exp. & Appl.* **19**:115–129.

63. Kaaya, G.P. and Darji, N., 1988, The humoral defence system in tsetse: Differences in response due to age, sex and antigen types, *Devel. Comp. Immunol.* **12**:255–268.

64. Kaaya, G.P., Ratcliffe, N.A., and Alemu, P., 1986, Cellular and humoral defenses of *Glossina* (Diptera: Glossinidae): reactions against bacteria, trypanosomes, and experimental implants, *J. Med. Entomol.* **23**:30–43.

65. Kaaya, G.P., Otieno, L.H., Darji, N., and Alemu, P. 1986, Defence reactions of *Glossina morsitans morsitans* against different species of bacteria and *Trypanosoma brucei brucei*, *Acta Trop. (Basel)*. **43**:31–42.

66. Kaminsky, R., Beaudoin, E., and Cunningham, I., 1988, Cultivation of the life cycle stages of *Trypanosoma brucei* spp, Acta Trop. (Basel). **45**:33–43.

67. Katzin, A.M. and Colli, W., 1983, Lectin receptors in *Trypanosoma cruzi* an N-acetyl-D-glucosamine-containing surface glycoprotein specific for the trypomastigote stage, *Biochim. et Biophys. Acta.* **727**:404–411.

68. Lackie, A.M., 1988, Immune mechanisms in insects, *Parasitology Today* **4**:98–105.

69. Langley, P.A., Hall, M.J.R., Felton, T., and Ceesay, M., 1988, Determining the age of tsetse flies, *Glossina* spp. (Diptera: Glossinidae): an appraisal of the pteridine fluorescence technique, *Bull. ent. Res.* **78**:387–395.

70. Lehane, M.J., 1976, Formation and histochemical structure of the peritrophic membrane in the stablefly, *Stomoxys calcitrans*, *J. Insect Physiol.* **22**:1551–1557.

71. Lehane, M.J. and Mail, T.S., 1985, Determining the age of male and female *Glossina morsitans morsitans* using a new technique, *Ecol. Entomol.* **10**:219–224.

72. Lloyd, Ll. and Johnson, W.B., 1924, The trypanosome infections of tsetse flies in Northern Nigeria and a new method of estimation, *Bull. Ent. Res.* **14**:265–288.

73. Lloyd, Ll. et al., 1924, Second report of the tsetse fly investigation in the Northern Provinces of Nigeria, *Bull. Ent. Res.* **15**:127.

74. Makumyaviri, A.M. et al., 1984, Caracterisation de la capacite vectorielle de *Glossina morsitans morsitans* (Diptera: Glossinidae) vis-a-vis de *Trypanosoma brucei brucei* EATRO 1125 (AnTAR 1), *Ann. Soc. Belge Med. Trop.* **64**:365–372.

75. Makumyaviri, A.M. et al., 1984, Capacite vectorielle du type sauvage et du mutant salmon de *Glossina morsitans morsitans* Westwood, 1850 (Diptera: Glossinidae) dans la transmission de *Trypanosoma brucei* Plimmer et Bradford, 1899, *Cah. O.R.S.T.O.M., ser. Ent. med. et Parasitol.* **22**:283–288.

76. Maudlin, I., 1982, Inheritance of susceptibility to *Trypanosoma congolense* infection in *Glossina morsitans*, *Ann. Trop. Med. Parasit.* **76**:225–227.

77. Maudlin, I. and Ellis, D.S., 1985, Association between intracellular rickettsial-like infections of midgut cells and susceptibility to trypanosome infection in *Glossina* spp, *Z Parasitenkd.* **71**:683–687.

78. Maudlin, I. and Dukes, P., 1985, Extrachromosomal inheritance of susceptibility to trypanosome infection in tsetse flies 1. Selection of

susceptible and refractory lines of *Glossina morsitans morsitans*, *Ann. Trop. Med. Parasit.* **79**:317–324.

79. Maudlin, I. and Welburn, S.C., 1987, Lectin mediated establishment of midgut infections of *Trypanosoma congolense* and *Trypanosoma brucei* in *Glossina morsitans*, *Trop. Med. Parasit.* **38**:167–170.

80. Maudlin, I. and Welburn, S.C., 1988, The role of lectins and trypanosome genotype in the maturation of midgut infections in *Glossina morsitans*, *Trop. Med. Parasit.* **39**:56–58.

81. Maudlin, I. and Welburn, S.C., 1988, Tsetse immunity and the transmission of trypanosomiasis, *Parasitology Today.* **4**:109–111.

82. Maudlin, I. and Welburn, S.C., 1989, A single trypanosome is sufficient to infect a tsetse fly, *Ann. Trop. Med. Parasit.* **83**:432–433.

82a. Maudlin, I., Welburn, S.C., and Milligan, P., Salivary gland infection: a sex-linked recessive character in tsetse? *Acta trap.* (Basel) In Press.

83. Maudlin, I., Kabayo, J.P., Flood, M.E.T., and Evans, D.A., 1984, Serum factors and the maturation of *Trypanosoma congolense* infections in *Glossina morsitans*, *Z Parasitenkd.* **70**:11–19.

84. Maudlin, I., Dukes, P., Luckins, A.G., and Hudson, K.M., 1986, Extrachromosomal inheritance of susceptibility to trypanosome infection in tsetse flies. II. Susceptibility of selected lines of *Glossina morsitans morsitans* to different stocks and species of trypanosome, *Ann. Trop. Med. Parasit.* **80**:97–105.

85. Mews, A.R., Langley, P.A., Pimley, R.W., and Flood, M.E.T., 1977, Large scale rearing of tsetse flies (*Glossina* spp.) in the absence of a living host, *Bull. Ent. Res.* **67**:119–128.

86. Moloo, S.K., 1981, Effects of maintaining *Glossina morsitans morsitans* on different hosts upon the vector's subsequent infection rates with pathogenic trypanosomes, *Acta Trop. (Basel).* **38**:125–136.

87. Moloo, S.K. and Kutuza, S.B., 1985, Survival and reproductive performance of female *Glossina morsitans morsitans* when maintained on livestock infected with Salivarian trypanosomes, *Ann. Trop. Med. Parasit.* **79**:223–224.

88. Moloo, S.K. and Kutuza, S.B., 1988, Comparative study on the infection rates of different laboratory strains of *Glossina* species by *Trypanosoma congolense*, *Med. Vet. Entomol.* **2**:253–257.

89. Moloo, S.K. and Shaw, M.K., 1989, Rickettsial infections of midgut cells are not associated with susceptibility of *Glossina morsitans centralis* to *Trypanosoma congolense* infection. *Acta Trop. (Basel)* **46**:223–227.

90. Moloo, S.K., Kutuza, S.B., and Desai, J., 1987, Comparative study on the infection rates of different Glossina species for East and West African *Trypanosoma vivax* stocks, *Parasitology.* **95**:537–542.

91. Moloo, S.K., Kutuza, S.B., and Desai, J., 1988, Infection rates in sterile males of *morsitans*, *palpalis* and *fusca* groups *Glossina* for pathogenic *Trypanosoma* species from East and West Africa, *Acta Trop. (Basel).* **45**:145–152.

92. Molyneux, D.H., 1977, Vector relationships in the Trypanosomatidae, *Adv. Parasitol.* **15**:1–82.

93. Mutharia, L.M. and Pearson, T.W., 1987, Surface carbohydrates of

procyclic forms of African trypanosomes studied using fluorescence activated cell sorter analysis and agglutination with lectins, *Mol. Biochem. Parasitol.* **23**:165–172.

94. Mwangelwa, M.I., Otieno, L.H., and Reid, G.D.F., 1987, Some barriers to *Trypanosoma congolense* development in *Glossina morsitans morsitans*, *Insect Sci. Applic.* **8**:33–37.

95. Nitcheman, S., 1988, Comparaison des longevites des glossines (*Glossina morsitans morsitans* Westwood, 1850) infectees par les trypanosomes (*Trypanosoma nannomonas congolense* Broden, 1904) et des glossines saines, *Ann. Parasitol. Hum. Comp.* **63**:163–164.

96. Otieno, L.H., Darji, N., Onyango, P., and Mpanga, E., 1983, Some observations on factors associated with the development of *Trypanosoma brucei brucei* infections in *Glossina morsitans morsitans*, *Acta Trop. (Basel).* **40**:113–120.

97. Page, W.A., 1972, The infection of *Glossina morsitans* Weid. by *Trypanosoma brucei* in relation to the parasitaemia in the mouse host, *Trop. Anim. Hlth Prod.* **4**:41–48.

98. Pell, P.E. and Southern, D.I., 1976, Effect of the coccidiostat, sulphaquinoxaline on symbiosis in the tsetse fly, *Glossina* species, *Microbios Lett.* **2**:203–211.

99. Pereira, M.E.A., Andrade, A.F.B., and Ribeiro, J.M.C., 1981, Lectins of distinct specificity in *Rhodnius prolixus* interact selectively with *Trypanosoma cruzi*, *Science* **211**:597–600.

100. Pereira, M.E.A., Loures, M.A., Villalta, F., and Andrade, A.F.B., 1980, Lectin receptors as markers for *Trypanosoma cruzi* developmental stages and a study of the interaction of wheat germ agglutinin with sialic acid residues on epimastigote cells, *J. Exp. Med.* **152**:1375–1392.

101. Phelps, R.J. and Vale, G.A., 1978, Studies on populations of *Glossina morsitans morsitans* and *G. pallidipes* (Diptera: Glossinidae) in Rhodesia, *J. Appl. Ecol.* **15**:743–760.

102. Pires, F.A., Da Silva, J.M., and Teles E Cunha, G., 1950, Posicao actual da tse-tse na area da Sitatonga circunscricao do Mossurize. Estudo comparitivo da infestacao por tripanossomas patogenicos nas *G. morsitans*, *G. pallidipes*, *G. brevipalpis* e *G. austeni*, *Mocambique* **62**:1–59.

103. Rener, J., Carter, R., Rosenberg, Y., and Miller, L.H., 1980, Anti-gamete monoclonal antibodies synergistically block transmission of malaria by preventing fertilization in the mosquito, *Proc. Natl. Acad. Sci. USA.* **77**:6797–6799.

104. Renwrantz, L. and Schottelius, J., 1977, Charakterisierung der oberflache von trypanosoma brucei EATRO 427 mit Lektinen, Protektinen and Blutgruppen-Antiseren, *Z Parasitenk.* **54**:139–147.

105. Richner, D., Brun, R., and Jenni, L., 1988, Production of metacyclic forms by cyclical transmission of West African *Trypanosoma (T.) brucei* isolates from man and animals, *Acta Trop. (Basel).* **45**:309–319.

106. Robertson, M., 1913, Notes on the life history of *Trypanosoma gambiense*, with a brief reference to the cycles of *Trypanosoma nanum* and *Trypanosoma pecorum* in *Glossina palpalis*, *Philos. Trans. B.* **203**:161–184.

107. Ross, C.A., 1987, *Trypanosoma congolense*: differentiation to metacyclic

trypanosomes in culture depends on the concentration of glutamine or proline, *Acta Trop. (Basel).* **44**:293–301.

108. Ruddall, K.M., 1963, The chitin/protein complexes of insect cuticles, *Adv. Insect Physiol.* **1**:257–314.

109. Ryan, L., Kupper, W., Croft, S.L., Molyneux, D.M., and Clair, M., 1982, Differences in rates of acquisition of trypanosome infections between *Glossina* species in the field, *Ann. Soc. Belge Med. Trop.* **62**:291–300.

110. Saunders, D.S., 1962, Age determination for female tsetse flies and the age composition of samples of *Glossina pallidipes* Aust., *G. palpalis fuscipes* Newst. and *G. brevipalpis* Newst, *Bull. Ent. Res.* **53**:579–595.

111. Tarimo, S.A., Golder, T.K., Dransfield, R.D., Chaudhury, M.F.B., and Brightwell, R. 1985, Preliminary observations on trypanosome infection rates in *Glossina pallidipes* and the factors affecting them at Ngruman, Kenya. *International Scientific Council for Trypanosomiasis Research and Control.* 18th Meeting, Harare (Zimbabwe) pp 276–283. Publication No. 113 O.A.U./S.T.R.C.

112. Turner, C.M.R., Barry, J.D., and Vickerman, K., 1988, Loss of variable antigen during transformation of *Trypanosoma brucei rhodesiense* from bloodstream to procyclic forms in the tsetse fly, *Parasitology Research* **74**:507–511.

113. van Hoof, L.M.J.J., 1947, Observations on trypanosomiasis in the belgian congo, *Trans. R. Soc. Trop. Med. Hyg.* **40**:728–761.

114. van Hoof, L.M.J.J., Henrard, C., and Peel, E., 1937, Influences modificatrices de la transmissibilite cyclique du *Trypanosoma gambiense* par *Glossina palpalis*, *Ann. Soc. Belge. Med. Trop.* **17**:385–440.

115. Vickerman, K., 1965, Polymorphism and mitochondrial activity in sleeping sickness trypanosomes, Nature **208**:762–766.

116. Vickerman, K., 1985, Developmental cycles and biology of pathogenic trypanosomes, *Brit. Med. Bull.* **41**:105–114.

117. Wallbanks, K.R., Ingram, G.A., and Molyneux, D.H., 1986, The agglutination of erythrocytes and Leishmania parasites by sandfly gut extracts: evidence for lectin activity, *Trop. Med. Parasit.* **37**:409–413.

118. Weitz, B., 1963, The feeding habits of *Glossina*, *Bull. Wld Hlth Org.* **28**:711–729.

119. Welburn, S.C. and Gibson, W.C., 1989, Cloning of a repetitive DNA from the Rickettsia-like organisms of tsetse flies (*Glossina* spp.), *Parasitology* **98**:81–84.

120. Welburn, S.C. and Maudlin, I., 1989, Lectin signalling of maturation of *T. congolense* infections in tsetse, *Med. Vet. Entomol.* **3**:141–145.

120a. Welburn, S.C. and Maudlin, I., 1990, Haemolymph lectin and the maturation of trypanosome infections in tsetse, *Med. Vet. Entomol.* **4**:43–48.

121. Welburn, S.C., Maudlin, I., and Ellis, D.S., 1987, *In vitro* cultivation of rickettsia-like-organisms from *Glossina* spp., *Ann. Trop. Med. Parasit.* **81**:331–335.

122. Welburn, S.C., Maudlin, I., and Ellis, D.S., 1989, Rate of trypanosome killing by lectins in midguts of different species and strains of *Glossina*, *Med. Vet. Entomol.* **3**:77–82.

123. Wigglesworth, V.B., 1929, Digestion in the tsetse-fly: A study of structure and function, *Parasitology.* **21**:288–321.

124. Wigglesworth, V.B., 1972, *The Principles of Insect Physiology*, 7th ed. London, Chapman and Hall, 827 p.

125. Wijers, D.J.B., 1958, Factors that may influence the infection rate of *Glossina palpalis* with *Trypanosoma gambiense* 1. The age of the fly at the time of the infected feed, *Ann. Trop. Med. Parasit.* **52**:385–390.

126. Wijers, D.J.B. and Willett, K.C., 1960, Factors that may influence the infection rate of *Glossina palpalis* with *Trypanosoma gambiense* II. The number and the morphology of the trypanosomes present in the blood of the host at the time of the infected feed, *Ann. Trop. Med. Parasit.* **54**:341–350.

127. Willett, K.C., 1966, Development of the peritrophic membrane in *Glossina* (tsetse flies) and its relation to infection with trypanosomes, *Expl Parasit.* **18**:290–295.

128. Yen, J.H., 1975, Transovarial transmission of *Rickettsia*-like microorganisms in mosquitoes, *Ann. N.Y. Acad. Sci.* **266**:152–161.

129. Yorke, W., Murgatroyd, F., and Hawking, F., 1933, The relation of polymorphic trypanosomes, developing in the gut of Glossina, to the peritrophic membrane, *Ann. Trop Med. Parasit.* **27**:347–350.

7
Mosquito Spiroplasmas

Claude Chastel and Ian Humphery-Smith

Introduction

Mosquitoes represent the most important group of disease vectors, transmitting to both domestic animals and human beings a variety of pathogens including viruses, rickettsias, protozoas, and helminths. They act as a vector when the female takes a bloodmeal from an infected host and subsequently from a susceptible vertebrate host.

Schematically, the mosquito is generally idealized as a flying machine, divided in compartments (midgut, hemolymph, muscles, central nervous system, genital tract, and salivary glands) clearly separated by anatomical and physiological (biological) "barriers."

In certain cases, such as myxomatosis of rabbits, the pathogen is transmitted mechanically, via the contamination of the insect's mouthparts by a highly resistant poxvirus. Here, there is no invasion of any internal compartment of the vector and the transmission depends upon a short interval between two blood meals.

Otherwise, the pathogen is generally transmitted biologically and may undergo a complex parasitic cycle or a multiple step replicative cycle, including a complicated "steeple-chase" to overcome a number of these biological barriers. Finally it reaches the salivary glands (or the mouthparts) of the mosquito and thus is ready for transmission to a susceptible host during a second bloodmeal.

Some pathogens, such as viruses, may reach the genital tract, whereby

Claude Chastel, Department de Microbiologie et Santé Publique, Faculté de Médecine, 29285 Brest, France.
Ian Humphery-Smith, Department de Microbiologie et Santé Publique, Faculté de Médecine, 29285 Brest, France.

the infection is transmitted vertically to eggs either through the ovaries or during the oviposition (117).

The global capacity of a particular mosquito species to be infected by and to transmit a pathogen in a defined area (its "vector competence") is often analyzed in terms of geographical strain variation or genetic factors, that is, "intrinsic" factors (66). Other factors, that is, the "extrinsic" factors, are acquired by the mosquito either congenitally (endogenous latent viruses) or by contact with the environment and include bacterioflora of the gut, microsporidia, fungi, rickettsia-like organisms, and a variety of ectoparasites (culicid mites, *Culicoïdes anophelis*). Although probably important, these latter factors are neglected generally.

Among extrinsic factors, mycoplasmas and the more recently identified spiroplasmas (Class Mollicutes) appear noteworthy. Spiroplasmas are wall-less prokaryotes that show helical morphology in their exponential growth phase and that have been found in association with numerous arthropods and plants (151). At least some spiroplasmas, when experimentally inoculated into vectors of arbovirus infections (*Aedes aegypti*) or of malaria (*Anopheles stephensi*) reduce significantly their life span (73). These organisms may also be capable of interfering with replicative cycles of viruses and perhaps parasites in different parts of the mosquito's body (7, 8, 95).

Ever since the first isolation of a mosquito spiroplasma (msp) by Slaff and Chen in 1982 (129) from *Aedes sollicitans* in the United States, it was postulated that msp could affect possibly the vector competence of this mosquito for equine encephalitis viruses. Other spiroplasmas have been isolated subsequently from several species of mosquitoes in Europe, North America, and the Far East (21, 33, 124). These "new" organisms deserve our attention because of: (1) the high prevalence in which they are found in natural populations of mosquitoes from different countries and biotopes; (2) their capacity to multiply rapidly to high titers (24 h) in both artificial media and insect cell cultures; (3) their potential as biological insecticides directed towards the control of mosquitoes of medical or welfare importance; (4) the fact they represent *a priori* excellent model(s) for studying the pathogenicity power of mycoplasmas as a whole; and (5) the threat that thermophilic species may pose to human and animal health, even though they have yet to be isolated from mosquitoes (see Mosquito spiroplasmas and vertebrate hosts p. 181).

Generally, it is considered that female mosquitoes are obligate blood-feeders. This being so, many observations also indicate that female mosquitoes belonging to the *Aedes, Culex, Anopheles*, and so forth, genera, are active nectar-feeders. This may explain, at least in part, by what means mosquitoes acquire spiroplasmas in nature, that is, from flowers.

These reasons, and many others, largely justify a review on what is known presently about these enigmatic organisms.

History

In the early 1980s, Slaff and Chen (129), impressed by the fact that spiroplasmas, a newly described group of Mollicutes, were apparently widespread in nature, infecting plants, insects, and ticks, tried to isolate these organisms from the salt marsh mosquito, *Aedes sollicitans*, and from flowering salt marsh plants in West Creek, New Jersey. They used pools of 20 to 30 mosquitoes and the nutrient medium of Liao and Chen (91), in an attempt to isolate spiroplasmas. From a pool of 30 female *Ae. sollicitans* caught on August 18, 1981, they succeeded in isolating a helical organism, the strain AES-1, serologically unrelated to any previously described spiroplasma. It now bears the name *Spiroplasma culicicola*, (78). No isolates were recorded from 11 different plant species sampled when flowering at the same time of year. The authors emphasized that this "isolation from *Ae. sollicitans* may indicate a new and potentially serious threat to humans" because another spiroplasma, the "Suckling Mouse Cataract Agent" (*S. mirum*) induced cataract and lethal encephalitis in juvenile rodents and lagomorphs. They also evoked the possible effect of this msp on the vector competence of *Ae. sollicitans* for Eastern equine encephalitis virus (129).

Unaware of these results, in part due to the relative confidential status of the journal used by Slaff and Chen (129), our group in France was investigating *Aedes* mosquitoes for msp. Specimens collected on July 5, 1983, from the Isère river valley, Savoia, and transferred to liquid nitrogen prior to transport to the virus laboratory in Brest, proved positive for six isolates of msp when grown in SP4 medium (21): one from a mixed pool of *Aedes sticticus/Ae. vexans* and five from pools of *Aedes cantans* group. The first isolate, Ar 1343, has been described since as *S. sabaudiense*, (2), whereas the others were found to belong to the *Cantharis* spiroplasma complex, as established during ecological studies by Chastel et al. 1987 (24), in Savoia.

Since 1984, informal results about msp have been exchanged among interested scientists in France and the United States. This lead Léon Rosen (Pasteur Institute, Paris, and University of Hawaii) to attempt msp isolations from pools of mosquitoes stored at −70°C since 1981 during epidemiological studies conducted on the Japanese encephalitis virus in Taiwan and Japan. As expected, a large number of isolates were obtained and provisionally referred to as Sp 1, Sp 2, Sp 7, Sp 22, Sp 34, and so on. These isolates were obtained from a variety of *Aedes*, *Armigeres*, *Culex* and *Anopheles* mosquitoes. Among these isolates, a strain isolated from *Culex tritaeniorhynchus*, CT-1 (= Sp 1), was selected as a new spiroplasma by Clark et al. (33) and subsequently described as *S. taiwanense* (3).

In 1987, Tully et al. (138) revised the current serological status of spiroplasmas and added a number of new serogroups (XII–XXIII) to the then recognized spiroplasma serogroups previously established by Whitcomb et al. 1983 (145) that listed only one msp, AES-1 (*S. culicicola*) of serogroup X). *S. sabaudiense* and *S. taiwanense* were assigned to serogroups XIII and XXII, respectively.

During the summer of 1985, 1,298 mosquitoes belonging to seven genera and 21 species, collected in Macon County, Alabama, were tested for msp. Four msp were isolated from two pools of *Aedes fulvus pallens* (AEF-1, AEF-2), one pool of *Anopheles punctipennis* (ANP), and one pool of *Culex nigripalpus* (CXN) by Shaikh et al. (124). At least two of these isolates, AEF-2 and ANP, appeared serologically related to the *Cantharis* spiroplasma complex. The two others have yet to be classified further.

Finally, in 1987, Chastel et al. (24) presented results on their ecological studies performed in Savoia from 1983 to 1985. These investigations have been extended since (25) and now include observations from the "Loire Atlantique" district, where other mosquito species, *Aedes detritus* and *Ae. caspius*, collected from maritime biotopes yielded *S. sabaudiense* and a number of *Cantharis* spiroplasma complex isolates (90).

Taxonomy

The term *spiroplasma* was first coined by Davis and Worley, (39) as a trivial name to describe uncultivable, helical microorganisms associated with a plant disease, the "corn stunt." When similar organisms, associated with another plant disease [88], the "citrus stubborn" were cultivated by Saglio et al. in 1971, this trivial name was subsequently elevated to the rank of a genus (129, 120).

Razin and Freundt, (110), defined *spiroplasmas* as "cells helical during logarithmic growth, with rotary, flexionary and translational motility; genome size, 1×10^9 daltons; sterols required for growth; possess a phosphoenolpyruvate phosphotransferase systems for glucose; reduced nicotinamide adenine dinucleotide (NADH) oxidase activity is located only in cytoplasm; unable to synthetize fatty acid from acetate."

The type species is *Spiroplasma citri* (120). Subsequently, the genus *Spiroplasma* was elevated to the status of family, *Spiroplasmataceae*, by Skripal in 1974 (127, 128), belonging to the order *Mycoplasmatales* in the class Mollicutes.

Taxonomic Position of the Spiroplasma Genus Among Mollicutes

Literally speaking, Mollicutes (Division: Tenericutes) form a class of prokaryotes exhibiting a distinctive soft (*mollis*) inner side (*cutis*) ensuring facility to deform and to filter through membranes of defined porosity. In the past, filtrability has led a number of these mollicutes being confused with viruses and spirochetes.

According to the recommendations of the "Subcommittee on the Taxonomy of Mollicutes of the International Committee of Systematic Bacteriology" (May 1988) summarized by Tully (139), the class Mollicutes is divided in three orders, four families and six genera; this classification is discussed below.

Order Mycoplasmatales

FAMILY MYCOPLASMATACEAE

Mycoplasma and Ureaplasma are characterized by a genome size of 5×10^8 daltons (the smallest recorded in prokaryotes), a G+C content of DNA averaging 23 to 41 and 27 to 30, respectively, and a requirement for cholesterol for growth. Ureaplasma are distinct in possessing an urease. These organisms infect warmblooded vertebrates.

FAMILY SPIROPLASMATACEAE

Spiroplasma regroup organisms with a genome size of 10^9, a G+C content of 25 to 31, require cholesterol and exhibit helical morphology during the logarithmic growth phase. They are found in association with insects, ticks, and plants. In 1988, Razin (112) considered that the family Spiroplasmataceae warranted elevation to the ordinal rank of Spiroplasmatales.

Order Acholeplasmatales

FAMILY ACHOLEPLASMATACEAE

Acholeplasma possess a genome size of 10^9 daltons, a G+C content of 27 to 36 and do not require cholesterol for growth. They inhabit animals, plants, and insects (45).

Order Anaeroplasmatales

FAMILY ANAEROPLASMATACEAE

Anaeroplasma and Asterolplasma are comprised of obligate anaerobes with a genome size of 10^9, a G+C content of 29 to 33, and 40, respectively. They are encountered in the rumen of artiodactyles. Anaeroplasma require cholesterol for growth but Asterolplasma do not.

Phylogeny

Because of their extreme simplicity and minute dimensions, mollicutes have been the subject of much speculation concerning their phylogenetic origin (113). It is only with the advent of molecular biology that meaningful progress has been achieved in this exciting area of research.

Initially, Wallace and Morowitz, (142) proposed an evolutionary scheme in which mollicutes were placed at the root of a prokaryote evolutionary tree. The mollicutes with a genome size of 5×10^8 daltons (Mycoplasma and Ureaplasma) were designated "protokaryotes," whereas those with a genome size of 10^9, which have evolved from the protokaryotes by doubling their genome, could be regarded as intermediates with respect to the wall-covered prokaryotes (Eubacteria).

More recently, mollicutes have been regarded as distant relatives of prokaryotes, either as a "degenerative" branch of several bacterial progenitors (polyphyletic origin), or derived from a single monophyletic branch, having arisen deep within the gram-positive part of the eubacterial tree (113, 154).

Similarly, Rogers et al. (116) constructed a mycoplasma evolutionary tree based on 5 S rRNA sequence data. Analysis of such sequences have shown that mollicutes form a coherent phylogenetic group, which along with *Clostridium innocuum*, arose as a branch of low G+C % gram-positive bacteria, near the lactobacilli and the streptococci. The initial event in mycoplasma phylogeny was the formation of the *Acholeplasma* branch. A subsequent radiation produced the *Spiroplasma* genus which appeared at the origin, by genomic reduction, of all other sterol-requiring mycoplasmas: *Mycoplasma*, *Ureaplasma*, and *Anaeroplasma*. This explication is in agreement with previous indications of Woese et al. (152).

J.M. Bové et al. (1989) (personnel communication), after having sequenced a number of genes from spiroplasmas and their viruses (the spiroplasmaviruses Sp V1 and Sp V4), completed the above data and confirmed that spiroplasmas appear to have emerged from gram-positive eubacteria with a low G+C % content (*Clostridium* spp.) by genomic reduction and regressive evolution. Spiroplasmas, and other mollicutes, thus appear as the simplest autoreplicative prokaryotic cells, not because they are very "primitive" organisms but rather as a result of genome size reduction.

In fact, when a phylogenetic tree is constructed for the mycoplasmas and gram-positive bacteria based upon 16 S rRNA sequence homology, mollicutes represent a group of organisms in a state of rapid evolution (153, 154).

Spiroplasma Taxonomy

In 1981, Tully and Whitcomb (136) considered the taxonomy of the spiroplasma genus to be in its infancy and this remains true today. Classification of spiroplasmas is based essentially on antigenic relationships demonstrated by serological methods such as growth inhibition test and the more specific "deformation" test of Williamson et al. 1978 (149). This later test is able to distinguish spiroplasma isolates and classify them in numbered serological "groups." The deformation test also allows "sub-groups" or serovars to be identified, as seen in the group I, which has already been divided into I–1, I–2, I–3, I–4, and so forth. Such antigenic differences have been demonstrated also by using the ELISA test (78, 121). The value of serological grouping and subgrouping was confirmed by Mouches et al. (103) who compared the protein profiles of spiroplasmas by one- and two-dimensional gel electrophoresis, and by Christiansen et al. (28) using DNA–DNA hybridation.

Several groups or subgroups of spiroplasmas have been ascribed "species" status now, for instance *Spiroplasma citri* for I–1, *S. melliferum* for I–2, *S. Kunkelii* for I–3, and so on. For many others, this is not the case yet. This may appear as a source of confusion for the noninitiated and is a situation aggravated by the fact that certain strains have changed their group number since isolation. Description of new species is needed urgently, as is seen for group XVI, the *Cantharis* spiroplasma, an ensemble of related serovars infecting mosquitoes and other insects. This is also true for the increasing number of isolates obtained from tabanids.

Described Species of Mosquito Spiroplasmas

There are only three fully described species of msp: *S. culicicola* (group X), *S. sabaudiense* (group XIII), and *S. taiwanense* (group XXII). (cf. Table 1).

S. culicicola, strain AES-1, was isolated first in August 1981, in New Jersey, from a pool of 30 female *Ae. sollicitans* by Slaff and Chen (129) and described by Hung et al. in 1987 (78). It is a monotypic representative of group X. Most cells of strain AES-1 are nonhelical and differ in serological reactivity, growth pattern, protein profiles, and morphology from pre-viously described spiroplasmas. Optimal growth has been reported at 31°C, whereas no growth was observed at 37°C. The G+C % content of DNA was approximately 26. Another strain in group X, BA-1, has been isolated from the gut of a red-spotted purple butterfly (*Limenitis arthemis astyanax*) in the United States.

S. sabaudiense, strain Ar-1343, was isolated first in July 1983 in the French northern Alps from a mixed pool of *Ae. sticticus/Ae. vexans*, by Chastel et al. (21), and described in 1987 by Abalain-Colloc et al. (2). It is a monotypic representative of group XIII. It exhibits helical morphology

TABLE 7.1. Described species of mosquito Spiroplasmas.

Species	Type strain	Host	Country	Serological group
Spiroplasma culicicola	AES-1[T]	*Aedes sollicitans*	New Jersey, U.S.A.	X
S. sabaudiense	Ar-1343[T]	*Aedes sticticus/ Ae. vexans*	Northern Alps, France	XIII
S. taiwanense	CT-1[T]	*Culex tritaeniorhynchus*	Taiwan, China	XXII

and motility typical of Spiroplasmataceae. Growth occurred at 20 to 32°C in artificial mediums. The G+C % content was found to be about 30.

S. taiwanense, strain CT-1, was recovered by L. Rosen from a pool of 100 female *Culex tritaeniorhynchus* collected in rice fields at Taishan, near Taipei, Taiwan, Republic of China, in July 1981 (33). It is a monotypic representative of group XXII and was described in 1988 by Abalain-Colloc et al. (3). This organism also exhibits the helical morphology and motility typical of spiroplasmas and grows at 22 to 30°C in artificial media. The G+C % content was found to be about 25.

These three msp all require cholesterol for growth. The type strains, AES-1^T (ATCC 35 112), Ar-1343^T (ATCC 43 303) and CT-1^T (ATCC 43 302) have been deposited in the American Type Culture Collection at Rockeville, Maryland, in the United States.

Species Under Evaluation

The *Cantharis* spiroplasma, group XVI, is not a monotypic species because it was isolated initially in the U.S.A. from different types of insects including beetles (*Cantharis carolinus*, *C. bilineatus*), a wasp (*Monobia quadridens*), the gut of a firefly (*Photinus pyralis*), and thereafter from many different species of mosquitoes in France and in the United States (33, 124). In addition, two isolates of group XVI spiroplasmas have been recovered from a flowering plant, the thistle (*Circium* spp.), in France (25).

A number of representative strains of the group XVI, Ar-1357 (mosquito), CC-1 (*C. carolinus*), and MQ-6 (*M. quadridens*) were found related but clearly dictinct when compared by reciprocal metabolism and growth inhibition tests (138). These differences were confirmed when these strains and others belonging to group XVI were compared by polyacrylamide gel electrophoresis for proteins and by DNA–DNA hybridization (M.L. Abalain-Colloc, 1989, personal communication).

Mosquito Spiroplasmas Insufficiently Characterized

From pools of Asian mosquitoes collected in Taiwan and Japan during epidemiological studies on Japanese encephalitis virus during 1981, Dr. Léon Rosen isolated a large number of spiroplasmas and nonhelical mycoplasmas that remain insufficiently characterized.

Results of investigations already carried out on these isolates by our laboratory or those of Dr. Joseph Tully (National Institute of Allergy and infections Diseases, Frederick, MD) and Dr. David Williamson (State University of New York, at Stony Brook, NY) are presented in Table 7.2. Additional work is needed to complete the identification of these msp.

TABLE 7.2. Mosquito spiroplasmas insufficiently characterized.

Strain	Host	Country	Provisional characterization
Sp 7	*Armigeres subalbatus*	Taiwan	Unrelated to groups 1–2, IV, XIII, XVI, and XXII
Sp 18 (CUAS-1)	*Culex annulus*	Taiwan	Related to MQ-1 (group VII)
Sp 19	*C. tritaeniorhynchus*	Taiwan	= Sp 18
Sp 22	*C. pipiens* (larvae)	Taiwan	Undefined *Acholeplasma* (*A. morum ?*)
Sp 34	*Aedes vexans*	Japan	Related to *S. melliferum* (group 1–2)
ANP	*Anopheles punctipennis*	Alabama	Related to group XVI
AEF-2	*Ae. fulvus pallens*	Alabama	Related to group XVI

Spiroplasmas from Other Hematophagous Arthropods

Msp are not the sole spiroplasmas isolated from hemotophagous arthropods. Isolations have been made from ticks predating those from mosquitoes, whereas more recently numerous isolates have been obtained from deerflies and horseflies in the United States (63) and in France (77).

Suckling mouse cataract agent (SMCA) was isolated from pooled extracts of the rabbit tick, *Haemaphysalis leporispalustris*, using embryonated hen's eggs in the early 1960s, as a part of an epidemiological study on Rocky Mountain Spotted fever in Georgia (29). This agent had been thought first to be a virus and it was only in 1974 that it was eventually recognized as a mycoplasma-like organism by Bastardo et al. (9) and as a spiroplasma in 1976, by Tully et al. (133). This spiroplasma is now known as *Spiroplasma mirum* and is a member of group V (137). It is the only spiroplasma experimentally pathogenic for a variety of vertebrate hosts, including chick embryos, juvenile mice, rats, syrian hamsters, and rabbits. Injected intracerebrally to baby rodents and lagomorphs, *S. mirum* induces progressive cataract and encephalitis. Other members of group V, also isolated from *H. leporispalustris* ticks in the U.S.A. include GT-48 and TP-2 strains.

Another isolate from *H. leporispalutris*, strain 277 F, was obtained by Pickens et al. (106) from Montana, and initially considered as a spirochete. In fact, it was recognized as a spiroplasma by Brinton and Burgdorfer, (18) and belongs to a quite different group, I–4, closely related to plant pathogens, such as *S. kunkelii*. This finding led to some doubt about the nature of the biological association of 277 F with the tick.

A further eight isolates obtained from extracts of *Ixodes pacificus* ticks in Oregon were found to belong to group VI (135, 136). The representative strain of these isolates, Y 32, is characterized by an unusual property among spiroplasmas; viz., the ability to hemadsorb guinea pig erythrocytes. All eight isolates exhibited predominantly nonhelical morphology.

Attempts made in our laboratory in Brest to isolate spiroplasmas from a large number of ticks (> 1,000) from the old world, belonging to five genera, extracted both in pools or individually have been unsuccessful (22).

Thus tick-associated spiroplasmas have been isolated from only two species of Nearctic ticks (*H. leporispalustris* and *Ixodes pacificus*), but that belong to three different groups: V, VI, and less conclusively I–4.

On the contrary, work conducted on tabanid flies has led to a rich harvest of new groups and presumably species. Among the partially characterized strains from the United States, it is already possible to distinguish a "deerfly spiroplasma" (group XVII) from *Chrysops* spp., and two "tabanid spiroplasmas" from *Tabanus nigrovittatus* (group XVIII) and *Tabanus gladiator* (Group XXIII) (138). Several other North American isolates may constitute new groups (XXV, XXVI, XXIX) but are still under evaluation.

During the summer months (1989) in western France, our group recovered some sixteen isolates from *Tabanus bromius*, *T. sudeticus*, *T. bovinus*, *Hematopota pluvialis*, *H. pellucens*, *Hybomitra bimaculata*, *Chrysops pictus*, and *C. caecutiens*. Unrelated to msp and bee spiroplasmas, these are also under evaluation (77).

Methods of Study

The intensive study of spiroplasmas was made possible only when these new prokaryotes could be propagated in artificial media. Prior to this, dark-field illumination, phase contrast and electron microscopy in tissues or extracts of diseased plants, showed only mycoplasma-like organisms.

The first spiroplasma to be cultivated in vitro was the etiological agent of citrus stubborn disease (119) (47). This spiroplasma (group I–1) is now known as *S. citri*, (120) and is the type species for the genus *Spiroplasma*.

A short time thereafter, another plant pathogen, the agent of corn stunt disease, already characterized by morphological and ultrastructural methods (38), was found to be readily cultivable in artificial media (27, 148). This agent (group I–3) has been described since as *S. kunkelii* (146).

Thus artificial media have ensured the isolation and characterization of an increasing number of spiroplasmas from ticks (previously confused with viruses or spirochetes), honey bees, plants, flowers, and miscellaneous insects including leafhoppers, mosquitoes, and so on. However, other "fastidious" spiroplasmas, such as the Sex-Ratio Organism of *Drosophila* remained unable to cultivate in vitro until the efforts of Hackett and Lynn (60) and Hackett et al. (61).

Isolation Procedures

All the msp presently isolated were obtained directly by inoculating filtered extracts from pools or individual mosquitoes into artificial mediums.

The first isolate obtained by Slaff and Chen (129) made use of a nutrient medium similar to that described by Liao and Chen that (91) was maintained at 30°C for 1 week. As did subsequent authors (21, 124), Slaff and Chen recovered their isolates from pooled mosquitoes, a practice commonly employed by arbovirologists when searching for viruses from hematophagous arthropods.

Chastel et al. (25) have been able to show that isolation attempts of msp from pooled mosquitoes to be less sensitive than equivalent assays carried out from individual mosquitoes. The "actual isolation rate" or AIR, that is, the true rate of isolation of *S. sabaudiense* from individual *Ae. stricticus* was found to be as high as 26.8%, whereas the "minimal isolation rate" previously determined from pooled mosquitoes (*Ae. stricticus/Ae. vexans*) from the same area was only 0.3% (21).

Using this approach, it was possible to distinguish spiroplasma or other mollicute isolates obtained from external washings from those obtained from wholebody homogenates. In our study, all msp isolated came from the mosquito's body (25).

Therefore, the individual testing of mosquitoes would appear to be the most efficient method to study ecology and population dynamics of msp in the field. The procedure adopted in our laboratory for msp isolation is discussed below.

Each (adult) mosquito, either caught in the field or collected after emergence from nymphs in the laboratory, is put individually into a plastic tube and held until processing in (1) dry ice for transportation from the collection site, or (2) at −70°C in the laboratory.

The spiroplasma growth medium (SP4) originally described by Tully et al. (134) for the isolation of tick spiroplasmas is used for both isolation and characterization of msp. For details concerning the composition of this medium see Appendix 7.1. For interested readers, information on the composition of other artificial media such as BSR and M1A, can be found in the review of Tully and Whitcomb (136).

For msp assay, the plastic tube is thawed and each mosquito is processed

---

APPENDIX 7.1: Medium SP4 for isolation of tick spiroplasmas.[a]

The base medium is prepared by adding 3.5 g mycoplasma broth base (BBL), 10 g tryptone, 5.3 g peptone, and 5 g glucose to 615 ml deionized water. The base medium is sterilized by autoclaving at 121°C for 15–20 min. The final pH should be 7.5. The following sterile supplements are then added to the base medium:

CMRL 1066 tissue culture supplement (10×) (Gibco), 50 ml;
25% fresh yeast extract (Microbiological Associates), 35 ml;
2% Yeastolate solution (Difco), 100 ml;
Fetal calf serum (Flow) heat-inactivated at 56°C for 1 h, 170 ml;
Stock penicillin solution (100,000 units/ml), 10 ml; and
0.1% aqueous phenol-red solution, 20 ml.

[a] From Ref. 134.

individually. First, an external washing is performed in 2 ml of SP4 medium and the washing product is removed and filtered through a 0.45 μm Millifore membrane. The same specimen is ground carefully then in a chilled mortar with 2 ml of SP4. The suspension is filtered at 0.45 μm.

Both filtered washing products and wholebody homogenates, undiluted and diluted at 1:10, are incubated at 30°C for 10 days in plastic culture tubes. Positive culture is confirmed then by darkfield examination showing the presence of helical, motile organisms, 2 to 4 μm or longer, roughly resembling a spirochete. Negative cultures are blindpassaged once.

An Alternative Isolation Procedure

The technique outlined above is extremely expensive with respect to use of disposable filters and time, whereas nonhelical or helical forms closely associated with host tissue are probably lost in filtration. Furthermore, the above method does not lend itself to multiple homogenation techniques (44) that allow hundreds of arthropods to be processed daily. The latter is extremely important in the study of arthropod vector populations, as low-incidence rates can be missed easily unless thousands of individuals have been examined.

For these reasons, Grulet et al. (58) set out to find a spiroplasma-specific medium (Spiromed). The results obtained thus far are presented in Table 7.3. In addition, a variety of rotting material, animal faces, soil, water, and generally nonsterile samples showed no medium contamination, whereas one spiroplasma isolation has been obtained from a small sample ($n = 5$) of field-caught tabanids. Further testing is still needed to evaluate the efficacy of this medium for large-scale sampling of field-caught material. The protocol for spiroplasma isolation using this medium appears in Appendix 7.2.

APPENDIX 7.2: Protocol for spiroplasma isolation using spiromed.
___

Arthropod homogenates, soil, water, and diverse samples were held at dilution of 1:10 in the following solution for 45 min:

100 ml MEM medium (10×) (Seromed);
33 ml sodium bicarbonate at 5.6%;
10 ml glutamine 200 mmol/l (Biomerieux);
10 ml nonessential amino acids (Seromed);
100 ml fetal calf serum (heat-inactivated at 56°C for 30 min);
1,000,000 U/l of Penicillin;
1,000,000 U/l of Colimycin;
25 mg Amphotericin;
500 mg Vancomycin;
500 mg Rifampycin; and
Made up to 1 liter with distilled water.

Centrifugation was conducted where necessary (within the 45 min) to sediment particulate matter and a further 1:10 and a 1:100 dilution was made into modified SP4 medium containing 500,000 U/l, and sodium hypochlorite at 0.01%.

TABLE 7.3. Results obtained with SPIROMED.

Strain/species	Group	Host (family)	Location	CCU[a]/ml in SP 4	CCU/ml in "SPIROMED"
A 56 S. melliferum	I-2	Apis mellifera (Hymenoptera)	France	10[b]	10
SP 34 S. melliferum	I-2	Aedes vexans (Culicidae)	Japan	8	8
B 31 S. apis	IV	Apis mellifera (Hymenoptera)	France	9	9
SP 18	VII	Culex annulus (Culicidae)	Taiwan	11	9
Ar 1343 S. sabaudiense	XIII	Aedes sp. (Culicidae)	France	9	9
CC 1	XVI	Culex carolinus (Culicidae)	U.S.A.	9	9
Ar 1357	XVI	Aedes cantans (Culicidae)	France	8	7
SP 2 S. taiwanense	XXII	Anopheles sinensis (Culicidae)	Taiwan	10	10
SP 7	—	Armigeres subalbatus (Culicidae)	Taiwan	10	9
SPT 44	—	Tabanus bromius (Tabanidae)	France	10	10
SPT 161	—	Hybomitra bimaculata (Tabanidae)	France	8	8

[a] Color change units.
[b] log 10.

Arthropod Cell Lines

Isolation and culture of spiroplasmas have also been carried out in a wide variety of arthropod cell lines. This work has been reviewed by Yunker et al. (156). Humphery-Smith et al. (71) were able to demonstrate a life history for *S. sabaudiense* in *Ae. albopictus* C6/36 cells, whereas for both *S. sabaudiense* and *S. taiwanense*, the use of this cell line enabled persistent forms to be described for the first time for a member of the Class Mollicutes (74). Although quite feasible, we have not used arthropod cell lines for msp isolation in our laboratory.

CHARACTERIZATION PROCEDURES

The first step in characterizing any new spiroplasma isolate is to undertake triple clonage of the isolate by plating out liquid culture on solid SP4 medium. At this time, a good precaution is to lyophilize the triple-cloned isolate. Using this triple-cloned stock, we may characterize the strain then by a combination of serological, morphological, physiological, and biochemical methods.

Three serological methods are available for antigenic analysis of spiroplasmas. They are applied currently to both msp classification and ecological studies, that is the metabolism-inhibition test that represents a classical method for the study of mollicutes, the spiroplasma deformation test designed by Williamson et al. (149) and the disk growth-inhibition test designed by Whitcomb et al. (144). These tests used in combination were recommended for the serotyping of spiroplasmas (144, 138) and were applied to msp of groups X, XIII, XVI, and XXII.

With the reciprocal metabolism-inhibition test, the heterologous reactions are generally weak or nonexistent, but with the reciprocal deformation or growth-inhibition tests, the heterologous reactions may be relatively strong. Cross-reactivity may prevent the classification of isolates. This was the case for some isolates from Alabama (124). A similar case is seen among group XVI spiroplasmas that were isolated from mosquitoes and other insects. The results of reciprocal metabolism inhibition and growth-inhibition tests recorded by Tully et al. (137) are presented in

TABLE 7.4. Reciprocal metabolism-inhibition and growth-inhibition tests with three spiroplasmas of group XVI.[a]

	Results with antiserum prepared to strain		
Antigen	Ar-1357	CC-1	MQ-6
Ar-1357 (mosquito)	39,000/6[b]	162/5	162/4
CC-1 (beetle)	486/5	39,000/5	486/5
MQ-6 (wasp)	1458/5	4374/5	39,000/5

[a] From Ref. 138.
[b] Reciprocal of metabolism inhibition titer/zones of growth inhibition (in millimeters).

Table 7.4. From these results, it is obvious that the three strains (one msp and two others) share only partial serological affinity. The metabolism-inhibition test proved more discriminative for separating group XVI serovars, whereas the disk growth-inhibition test was essentially a "group specific" test.

A problem associated with the serological characterization of msp isolates is the difficulty in obtaining a clear-cut classification of isolates in either serovar or species. For group I, this problem was solved by defining an equivalence between species and certain subgroups: I–1 = S. citri; I–2 = S. melliferum; I–3 = S. kunkelii; ... I–8 = S. phoeniceum, but this is not true for I–4, I–5, and I–6. For group XVI, no serovar has been described as a species and decisions from spiroplasma taxonomists are needed urgently.

Among morphological methods, dark-field microscopy (magnification of 1250X) is very convenient for screening the isolates in logarithmic growth phase. It is used also in the deformation test to appreciate the morphological alterations induced by antibodies. However, a more complete characterization of a spiroplasma, for instance, a "new" strain or serovar, requires access to a transmission electron microscope.

The negative-staining technique (70) is a simple, rapid, and very useful tool for confirming the helical morphology, and for determining the diameter of the organism, about 100 to 200 nm (Fig. 7.1b). This method essentially rules out any confusion with spirochetes or Borrelia, which possess a peptidoglycan cell wall and periplasmic flagellae never encountered in spiroplasmas. Electron microscopic studies of ultrathin sections provide information concerning the outer limiting unit membrane of spiroplasmas (Fig. 7.1a). Organisms are grown in approximately 20 ml of liquid medium and pelleted by centrifugation (1500 rpm for 10 min). The cells are fixed then for 2 h in 3% glutaraldehyde, postfixed in 1% osmium tetroxyde for 1 h, dehydrated in acetone, embedded in Epon or Epon-Araldite, sectioned and stained with 1% aqueous uranyl acetate and Reynold's lead citrate. The cell wall of S. sabaudiense and S. taiwanense have been shown to be a unit membrane (2, 3, 21).

Physiological and biochemical methods include the determination of temperature requirements for growth, cholesterol and amino acid requirements, and several tests for the ability of the strain in question to catabolize glucose or other carbohydrates in defined medium and to hydrolyze arginine and urea. The genomic analysis is limited generally to DNA extraction and determination of guanine plus cytosine (G+C) content by procedures that depend on both buoyant density and melting temperature. Further analysis may include DNA–DNA hybridization (28) or DNA–rRNA hybridization (139). Uni- and bidimensional electrophoresis of proteins may also be a useful adjunct, as, for example, in comparing profiles of serovars of the group XVI spiroplasmas (M.L. Abalain-Colloc, 1989, personal communication).

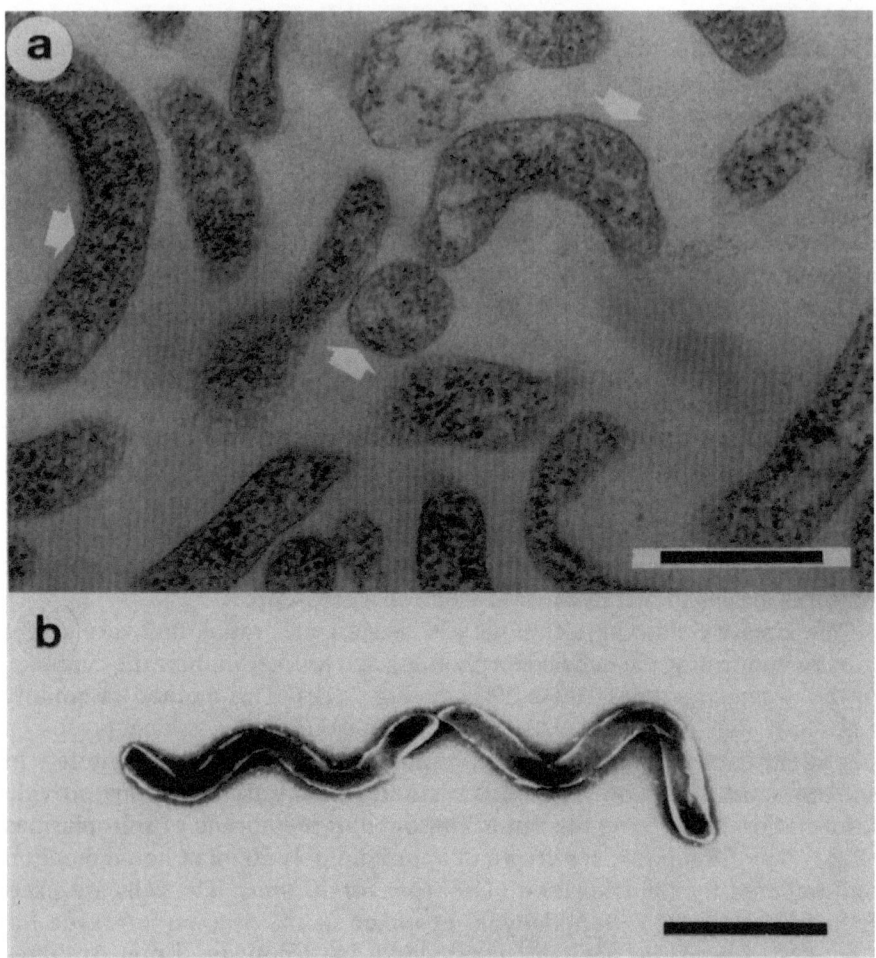

FIGURE 7.1. Electron micrographs of *Spiroplasma sabaudiense* showing (a) the unit membrane (arrow); and (b) transverse binary fission producing two elementary helices, as yet not fully separated. Scale bar (a) 0.3 μm; (b) 1.0 μm.

Immunofluorescent and ELISA assays have been used in our laboratory also to study msp utilizing an antibody directed against surface membrane proteins and prepared following the method of Wroblewski et al. (155).

Other technologies such as monoclonal antibodies, immunogold staining, or the development of sensitive molecular probes for assaying the msp in insects and flowers' tissues or in surface waters would improve greatly our ability to acquire knowledge of the ecology and pathogenicity of these organisms.

Biological Properties of Msp

The three msp already described, *S. culicicola* (strain AES-1), *S. sabaudiense* (Strain Ar 1343), and *S. taiwanense* (strain CT-1) have been characterized according to minimal criteria laid down by the International Committee on Systematic Bacteriology Subcommittee on Taxinomy of Mollicutes, 1979 (79). Thus the original description of these species (2, 3, 78) were accompanied by many basic data on their morphological and biological properties. Most of these characters are shared by other spiroplasmas. To date only scanty information is available concerning the *Cantharis* spiroplasma complex and, of course, those msp that have been characterized insufficiently (see above).

Morphological Properties

The three species of msp satisfy fully the morphological criteria required for inclusion in the genus *Spiroplasma* (34).

Liquid cultures examined by dark-field microscopy and electron microscopy show helical cells entirely devoid of a cell wall or periplasmic fibrils, with a diameter of about 100 to 200 nm. These cells are 1 to 2 μm long, exhibiting rarely more than one or two turns, for *S. culicicola*, and 3.1 to 3.8 μm long for the two other species. Thin sections show a triple-layered unit membrane surounding the cell cytoplasm.

On solid media, *S. culicicola* and *S. taiwanense* form "fried-egg" colonies typical of mollicutes, whereas *S. sabaudiense* colonies are diffuse and accompanied by satellite colonies.

Spiroplasmas have been shown to be highly pleomorphic to such an extent that it is difficult to imagine an association between structure and function for all the stages thus far described. Nonetheless, some order is becoming apparent among this seemingly endless morphological diversity.

As indicated by the term, *spiroplasma* (39), it was helical filaments that first drew attention to the differences between these enigmatic organisms and mycoplasmas (38, 88, 119). Fudl-Allah (47) using *Spiroplasma citri* (Group I), Garnier et al. (51, 52, 150) and, Bové et al. (16) demonstrated the role of helical forms in exponential growth and reproduction, and described elementary helices (the products of this reproductive process in a cell-free medium). These authors produced evidence that culture viability in a cell-free medium depended upon the presence of helical forms. The formation of round "spore-like bodies" or "blebs" (cf. 34) was observed in aging cultures and thought to represent degenerative forms under suboptimal conditions.

Cytadsorption of helical forms to arthropod cells in vitro is associated with a metamorphosis into a "coccoid" form. (51, 70, 71). Similar forms have also been observed intracellularly (cf. 63, for review), whereas Garnier et al. (53) showed these forms to be capable of intracellular reproduction. However, at least in some cases, forms contained within cytolysosomes would appear to be undergoing degradation (31).

A number of bizarre forms including "microcolonies" or "medusa heads" have been reported and would seem also to be associated with aging cultures (21, 51). In addition, budding and distorted forms have been reported (47). However, prior to the recent observations of Humphery-Smith et al. (76), much of this material can probably be taken to represent poorly fixed material.

Persistent forms have been shown to exist and will be discussed later. An example of the morphological diversity seen among msp, and in particular *S. sabaudiense*, is presented in Figures 7.2 through 7.5.

Biological Properties

All the three msp pass through membrane filters with pore diameters of 450, 300, and 200 nm but not those of 100 nm porosity.

Optimal growth occurs at 31°C for *S. culicicola* and in the range of 22 to 30°C for *S. sabaudiense* and *S. taiwanense*. None of the three msp is able to grow at 37°C, a very important finding which means that infection of warm-blooded vertebrates is highly unlikely.

All three msp require cholesterol to grow and ferment glucose, as do other spiroplasmas. Using the chemically defined medium CC–494, Hung et al. (78) showed that *S. culicicola* utilize fructose, mannose, mannitol, and trehalose. None of these organisms hydrolyse urea. Arginine is catabolized by *S. sabaudiense* but not by the two other msp. Pollack et al. (107) assayed some 67 enzyme activities for cell-free extracts of 10 spiroplasmas species including *S. sabaudiense* and *S. culicicola*. Unlike other spiroplasmas, detectable levels of dCMP kinase were absent for *S. culicicola* and thus this organism may not salvage deoxycytidine for deoxycytidine triphosphate synthesis and, as opposed to earlier observations (96), no uridine phosphorylase activity was detected. As with other spiroplasmas, these two msp were principally fermentative, possessing enzyme activities that converted glucose 6-phosphate to pyruvate and lactate by the Embden–Meyerhof–Parnas pathway. They were also capable of substrate phosphorylation and synthesis of purine mononucleotides by using pyrophosphate as the orthophosphate donor, and by using adenosine triphosphate (ATP) to phosphorylate deoxyguanosine, in contrast to other mollicutes. All spiroplasma spp. strains had deoxyuridine triphosphatase activity. As with all other mollicutes, a tricarboxylic acid cycle was apparently absent, whereas reduced nicotinamide adenine dinucleotide oxidase activity was localized in the cytoplasmic fraction of all spiroplasma species tested.

Neither *S. sabaudiense* nor *S. taiwanense* (colonies on agar) hemadsorb guinea pig erythrocytes, a property apparently restricted to the tick-associated spiroplasma of group VI (see above). They are also incapable of

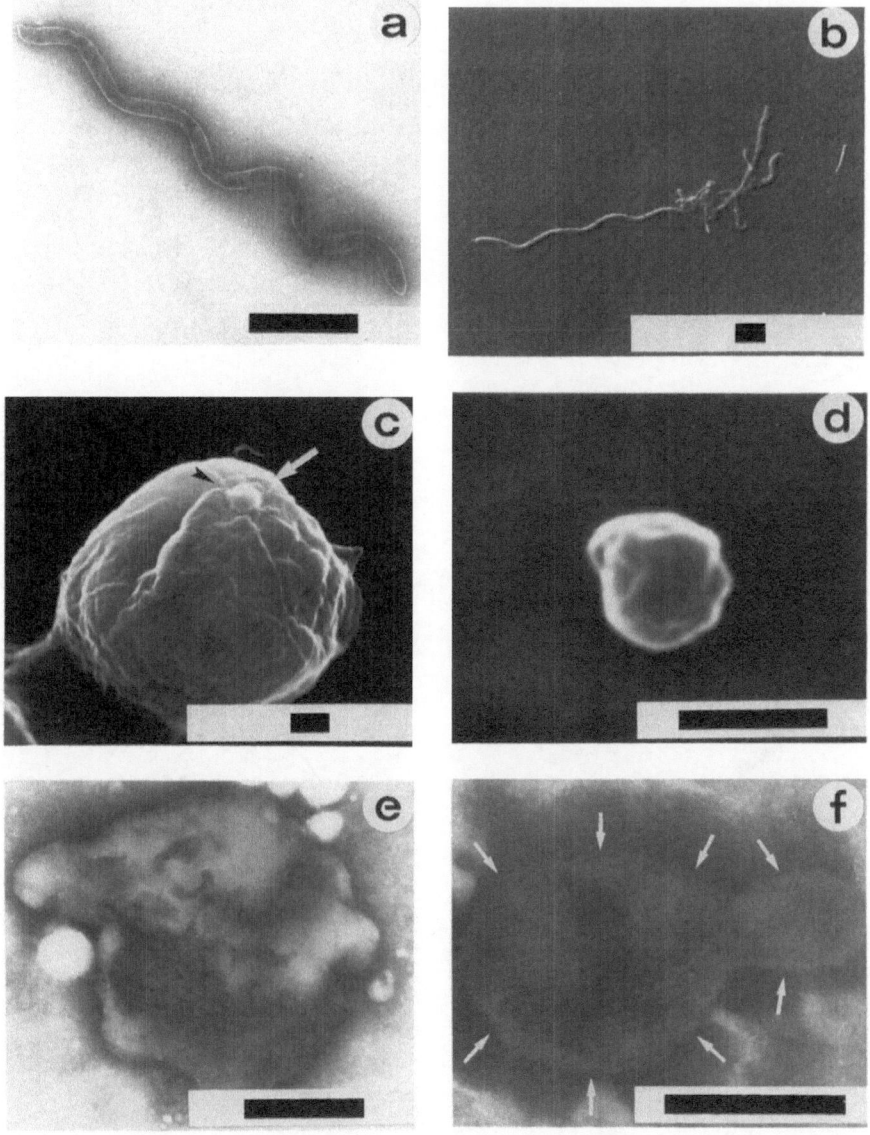

FIGURE 7.2. A variety of life-history forms of *Spiroplasma sabaudiense*. (a) exponential growth phase showing helical morphology; (b) microcolony; (c) cytadsorption of a mosquito cell concomitant with morphogenesis to a coccoid form; (d) coccoid form; (e) degenerate form or "bleb"; (f) persistent form or "ampullae" entrapped within cellular debris. Scale bar (a), (b), (c), (d), 1.0 μm; (e), (f), 0.4 μm.

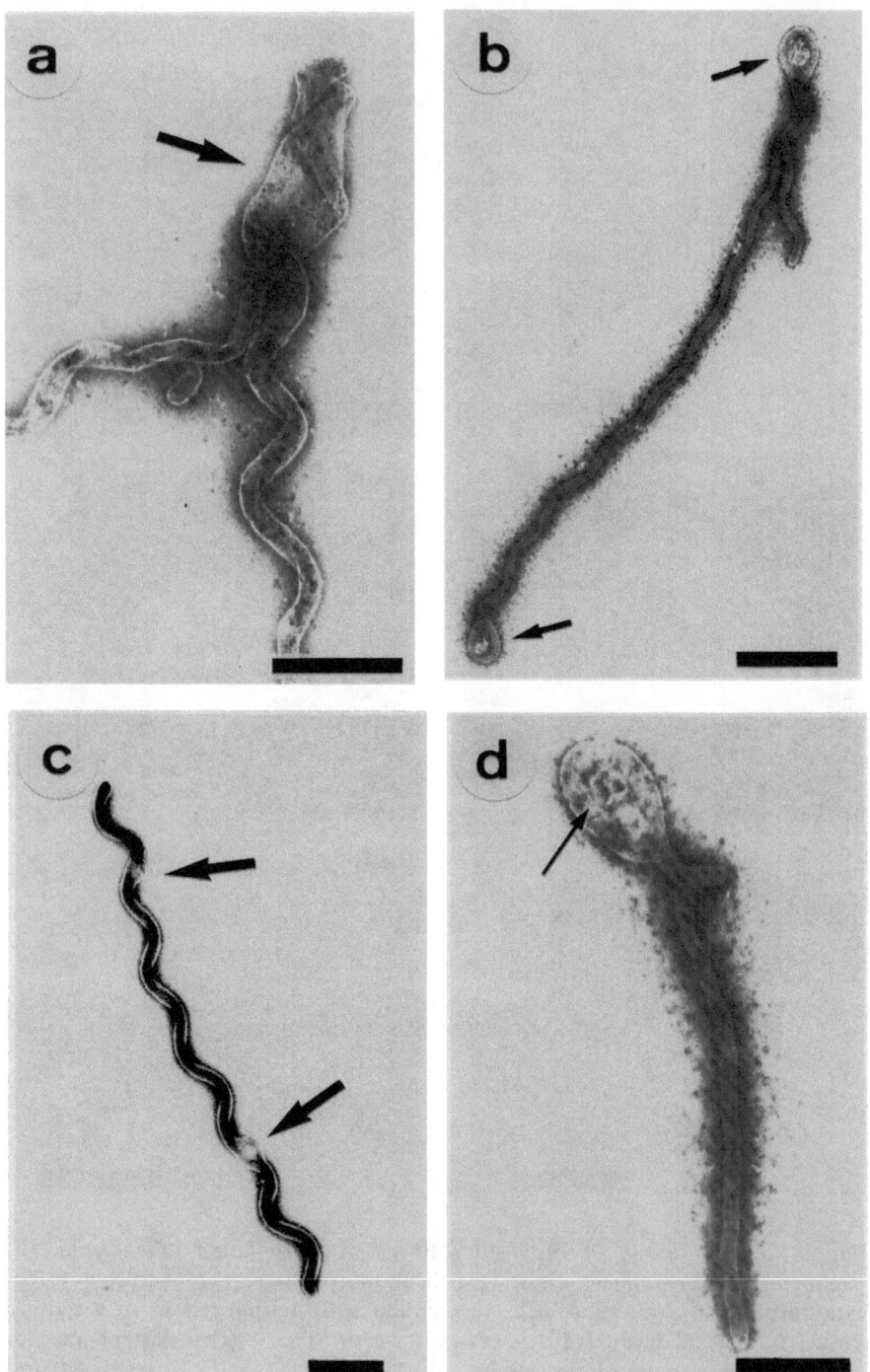

FIGURE 7.3. *Spiroplasma sabaudiense* at 72 h postinoculation in the supernatant of *Aedes albopictus* C6/36 cell cultures showing the rounding off of helical forms by (a) apparent fusion of the plasma membrane (arrow) or (c) swelling of the plasma membrane (arrow); (b) and (d) represent examples of this "rounding off" to produce swellings containing "embryo-like" structures or "spiroplasmagenic corpuscles." Cultures containing these forms are capable of returning to exponential growth phase when subcultured into fresh media. Scale bars (a), (c), (d) 1 μm; (b) 2 um.

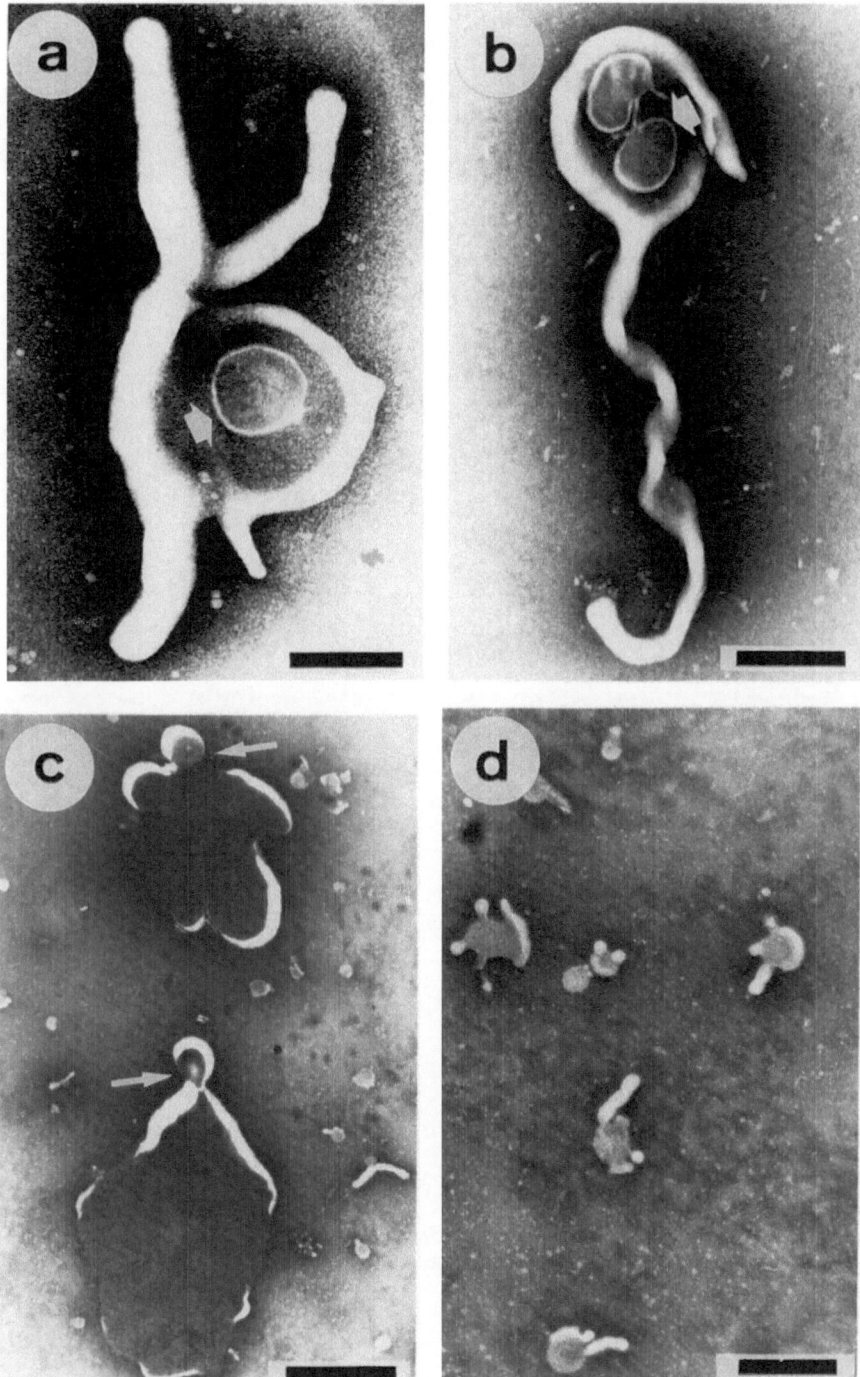

FIGURE 7.4. Cultures of *Spiroplasma sabaudiense* stressed by centrifugation for 20 min (a, b) and 60 min (c). Following centrifugation cultures remain viable and in (a) and (b) "spiroplasmagenic corpuscles" (arrow) are again visible. Following centrifugation for 1–5 h, helical morphology is modified uniformly into "balloon-like" structures. When filtered at 0.2 μm, these "balloons" are theoretically too large to filter and thus viability of filtrates must depend upon either the "bud-like" structures (arrow) in (c) or the debris seen in (d). Scale bars (a), (d) 0.3 μm; (b) 0.5 μm; (c) 0.8 μm.

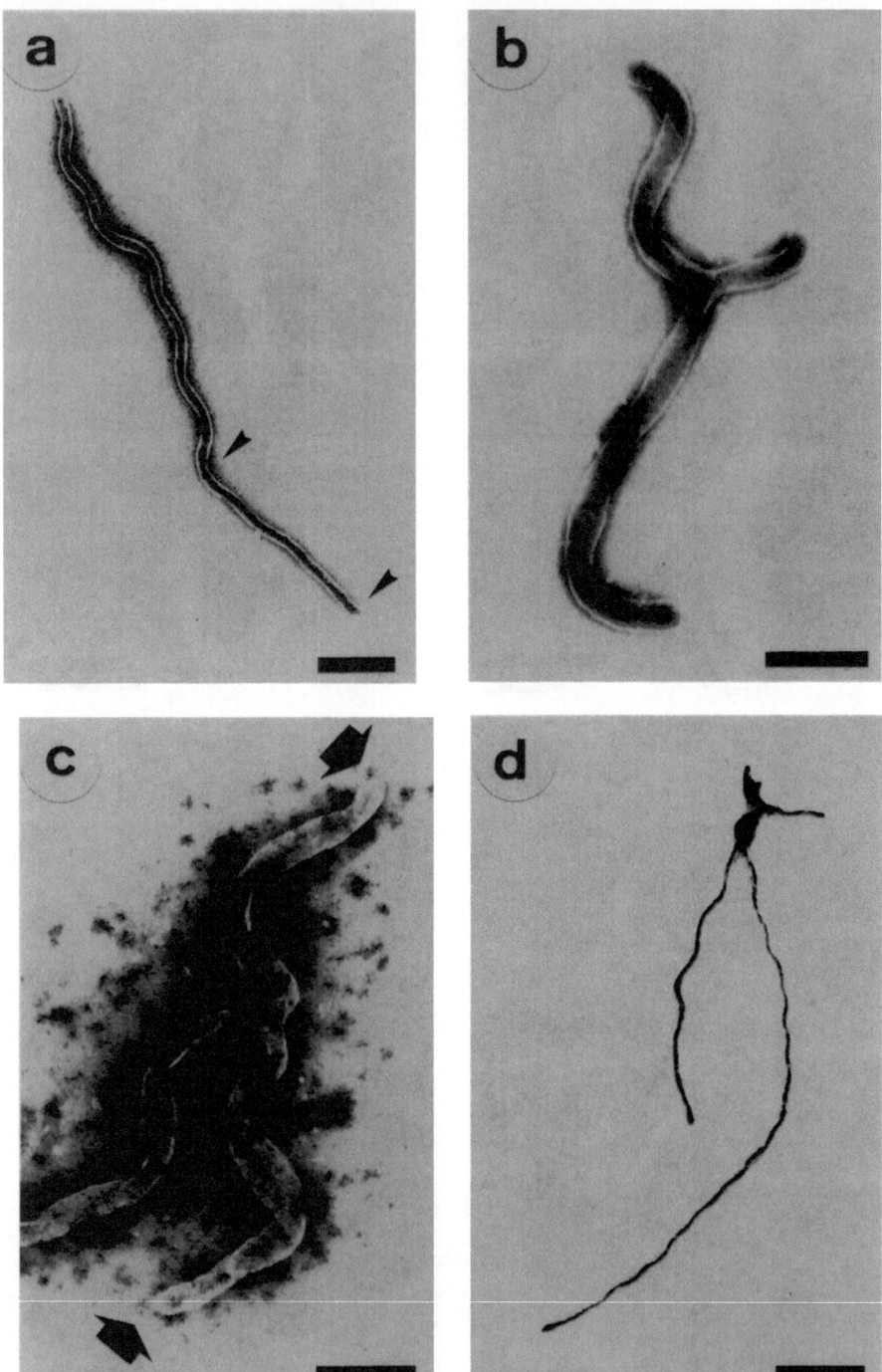

FIGURE 7.5. In postexponential growth phase (72 h), (a) helicity of *Spiroplasma sabaudiense* is lost (arrowed) and longitudinal binary fission (b, c) can be observed, however, it is not known if the latter represent true replicative stages or degenerative forms. (c) Tip regions of a helical form possessing 12 spirals are arrowed; (d) Branching and microcolony formation can be observed also in aging colonies. These forms do not remain viable for extended periods at 28 or 37°C in non-cell-associated culture and disappear progressively in cell-associated culture. Scale bar (a) 1.0 μm; (b, c) 0.5 μm; (d) 2 μm.

hemadsorbing human, sheep, chicken, or mouse blood at 4, 20, and 37°C. Similar results were obtained for hemacytadsorption (infected C6/36 cells) at 20°C. Freon-treated antigens of these two msp did not show evidence of hemaglutinins at 4, 20, and 37°C at pH 5.8, 6.0, 6.2, 6.3, 6.4, 6.5, 6.6, and 6.8 (I. Humphery-Smith, O. Grulet and M. Mellouet, unpublished).

The G+C content of the DNA determined by both melting temperature and from their buoyant density are presented in Table 7.5. These values are in agreement with those found among other characterized spiroplasmas, that is, 24 to 30% (138).

Susceptibility to Antibiotics and Disinfectants

As for other mollicutes, spiroplasmas lack a peptidoglycan cell wall and are thereby naturally resistant to penicillin. Accordingly, the media used for primary isolations of spiroplasmas from ticks or other blood-sucking arthropods, such as SP4 medium, contain important concentrations of penicillin in order to prevent undesirable bacterial contaminations.

Spiroplasmas are also naturally resistant to rifampicin, an inhibitor of transcription. As underlined by Bové, 1989 (J.M. Bové personal communication), *Clostridium innocuum* and *C. ramosum* are also insensitive to rifampicin and thus mollicutes may have inherited the rifampicin insensitivity from their eubacterial ancestors. The rifampicin insensitivity, the low number of ribosomal RNA operons, and the use of UGA as a tryptophane codon, are specific to spiroplasmas.

Mollicutes including spiroplasmas has been found highly susceptible to tetracyclin, an antibiotic that inhibits protein synthesis at the ribosomal level. Accordingly, plants infected with spiroplasmas or mycoplasma-like organisms (MLO) can be treated with this antibiotic. The suggestion that antibiotic-resistant strains of spiroplasmas might emerge in diseased plants repeatedly treated by tetracyclin has stimulated studies on the susceptibility to antibiotics not only of plants pathogens such as *S. citri* and *S. kunkelii*, but also of honey bee pathogens and the tick-derived *S. mirum*.

All the strains of spiroplasmas tested thus far have been sensitive to tetracyclin, tetracyclin HCl, erythromycin, carbomycin, and tylosin. Without any exception, they were highly resistant to penicillin, ampicillin, sulfanilamide, bacitracin, and rifampicin. Large strain-variations were observed in their susceptibility to gentamycin, tobramycin, kanamycin,

TABLE 7.5. Guanine plus cytosine content.[a]

Msp species	Melting temperature	Buoyant density
S. culicicola	26.0	26.7
S. sabaudiense	29.6	30.2
S. taiwanense	25.7	25.4

[a] As determined by melting temperature, and buoyant density.

neomycin, paronomycin, chloramphemicol, and so on (17, 20, 40, 92, 84).

This type of antibiotic spectrum was found when Abalain-Colloc et al. (1) tested the susceptibility to antibiotics of six strains of msp isolated from *Aedes* mosquitoes in France, including the type strains Ar 1343 of *S. sabaudiense* and Ar 1357 of the *Cantharis* spiroplasma. These authors found that each strain was highly susceptible to tetracyclin, oxytetracyclin, doxycyclin, erythromycin, chloramphenicol, and pefloxacin. At the same time, these strains exhibited resistance to rifampicin and variable degrees of susceptibility to kanamycin and gentamycin. Thus msp did not differ in their basic antibiotic spectra from other spiroplasmas.

Few studies have been devoted to the susceptibility of spiroplasmas to current disinfectants and heavy-metal salts. Certain spiroplasmas, such as the B31 strain of *S. apis* or the Sp 2 strain of *S. taiwanense*, appear as efficient contaminants in laboratories where mollicutes or tissue cultures are handled. This question is not a purely academic one, as claims made by Bastian (10) that spiroplasmas might be responsible for such central nervous system disorders in man as the Creutzfeldt–Jakob disease, and those made by Kotani et al. (85) that *S. mirum* transforms certain types of cell cultures, both mean great care is needed when handling these organisms in the laboratory. Thus the use of laminar-flow hoods, gloves, mask, and efficient surface disinfectants are considered essential.

Stanek et al. (130) tested the efficacy of four different disinfectants, (ethanol, formalin, glutaraldehyde, and phenol) in several concentrations, on spiroplasmas pathogenic for plants (*S. citri*), honey bee (*S. melliferum*, *S. floricola*), and on two tick-derived spiroplasmas (*S. mirum* and the strain 277 F). Their results showed that spiroplasmas displayed considerable resistance to these disinfectants in comparison with two eubacteria. Particularly, *S. melliferum* (strain BC-3) exhibited the highest resistance, whereas *S. citri* and tick-derived spiroplasmas showed a susceptibility comparable to that of *Escherichia coli* and *Staphylococcus aureus*. Surprisingly, BC-3 strain tolerated concentrations of 40% ethanol far in excess of the concentrations supported by *S. aureus*.

Whitmore et al. (147) studied in vitro the susceptibility of several strains of *S. citri*, *S. floricola*, *S. melliferum*, and a strain of group V spiroplasma to different heavy-metal salts. These strains were found most susceptible to mercuric chloride, silver nitrate, and cadmium sulfate, and least suscepti- ble to nickel chloride and zinc sulfate. *S. citri* (strains Maroc R8A2 and C 189) was the most suceptible species to five of eight heavy-metal salts, and *S. melliferum* (AS 576) and *S. floricola* (23–6) were generally the least susceptible. Little or no data are available on heavy-metal salt or laboratory-disinfectant susceptibility of msp. Of note is the ability of msp to grow in 0.01% sodium hypochloride solution (cf. Appendix 7.2). This is the dilution recommended for sterilization of baby's bottles, for example, whereas some msp have been observed to replicate in 1% sodium hypochloride in SP 4 medium [I. Humphery-Smith and O. Grulet, unpublished).

Ecology of Msp

Very little ecological data are available for the three species of msp fully characterized and this is particularly true for *S. culicicola* and *S. taiwanense*.

S. culicicola was isolated during mid-August, 1981, from a pool of salt marsh mosquitoes *Aedes sollitans*, in New Jersey. At the time of this discovery, Slaff and Chen (129) put forward a number of questions that are equally applicable to all msp: (1) Could mosquitoes be involved in transmitting spiroplasmas to plants or animals? (2) Could the presence of spiroplasmas in mosquitoes interfere with their vector competence for viruses? (3) Could mosquitoes acquire their spiroplasma infection from nectar as mosquitoes, including females, have frequent contacts with plants and flowers for nectar? In fact, none of these questions have been resolved completely.

For instance, Slaff and Chen (129) failed to isolate msp from a number of flowering plants collected from the marshy biotopes that yielded *S. culicicola* from *Ae. sollicitans* and suspected of being nectar sources for mosquitoes. The plants assayed and found negative for spiroplasmas were as follows: *Juncus girardi* (black-grass), *Verbana hastata* (blue vervain), *Oenothera fructicosa* (sundrop), *Sabatia stellaris* (marsh pink), *Ascelpias incarnata* (swamp milkweed), *Tencrium littorale* (germander), *Hibiscus palustris* (swamp rose mallow), *Kosteletzkya virginica* (seashore mallow), *Pluchea purpurascens* var. *suvculenta* (salt marsh fleabane), *Solidago sempervirens* (seaside golden rod), and *Baccharis lamifolia* (groundsel tree). However, according to Williamson et al. (151), another insect host for *S. culicicola*, a red-spotted purple butterfly (*Limenitis artemis astyanax*), was identified also in 1983 by Clark. This observation possibly indicates that flower-foraging insects, such as butterflies, may represent a source for mosquito contamination by *S. culicicola*.

When describing *S. culicicola*, Hung et al. (78) insisted on the necessity to consider the behavioral changes of *Ae. sollicitans* during its adult life in order to interpret the ecology of this msp. Salt marsh mosquitoes may travel long distances during the spring, obtaining food from various flowers, but during summer, the period where *S. culicicola* was isolated, their activities are confined primarily to salt mashes.

S. taiwanense was isolated repeatedly in July 1981 from *Culex tritaeniorynchus* and *Anopheles sinensis* mosquitoes caught from animal bait near rice fields in Taiwan. These mosquitoes were given 4 days at laboratory ambient temperature to digest their bloodmeals and then they were conserved at $-70°C$. Other ecological data are lacking as these mosquitoes had been collected primarily during epidemiological surveys on Japanese encephalitis virus, and not for msp studies.

Elsewhere, most data on msp ecology has been obtained from France during field studies carried out since 1983 in the northern Alps, Savoia, France, and in 1988 from Atlantic biotopes. (21, 24, 25, 90).

Study Areas

In the northern Alps, mosquitoes were collected from the junction of the Isère and Arc Rivers, in the middle of the Isère River Valley and from the mountain forest station, Jarrier (1538 m), near St-Jean-de-Maurienne, some 30 km to the south. River valley biotopes that yielded mosquitoes corresponded to three lowland stations (290–310 m) located in the vicinity of Grésy-sur-Isère and Aiton and consisted of occluded loops of the river temporarily flooded. Mountain forest biotopes favorable to mosquito larvae comprise ground inundated by thawing of snow.

During spring and summer (1983–1988), mosquitoes were collected as adults, either by aspiration from human bait, or by using an entomological sweep net in vegetation, or by collecting imagos with "hood nets" as they emerged. A small number of larvae was collected also.

During the period from 1983 to 1984, no attempt was made to fully differentiate the species of *Aedes sticticus*/*Ae. vexans*, *Ae. cantans*/*Ae. annulipes*, and *Ae. cinereus*/*Ae. geminus* females and thus collections of 25 to 30 specimens were pooled for storage at −70°C.

During the period from 1985 to 1988, mosquitoes were collected, identified, and then preserved individually, thus enabling accurate identification of the mosquito host when positive for a msp.

In western France ("Loire Atlantique" district), near La Baule and Nantes cities, mosquitoes were collected from May to September 1988, either by aspiration from human or animal baits or by using entomological sweep nets in vegetation. Altitude of Atlantic biotopes was generally low (8–12 m). All mosquito specimens were identified and preserved individually prior to msp testing.

Mosquito Species Infected by Msp in France

Up until now, only two species of msp, *S. sabaudiense* (group XIII) and the *Cantharis* spiroplasma (group XVI) have been isolated from both Savoia and Loire Atlantique, despite the fact that the biotopes concerned are very different (alpine valleys and Atlantic salt marshes).

As recorded in Table 7.6, five species of mosquitoes (*Aedes cantans*, *Ae. cinereus*, *Ae. rusticus*, *Ae. sticticus*, and *Coquillettidia richiardii*) in Savoia and two species (*Ae. caspius* and *Ae. detritus*) in Loire Atlantique were found infected by the *Cantharis* spiroplasma. On the other hand, hosts of *S. sabaudiense* appeared more restricted because only *Ae. sticticus* in Savoia and *Ae. detritus* in Loire Atlantique yielded this msp.

From studies conducted in Savoia, *Ae. sticitus* appeared as the species most heavily infected by msp: 26.8% of females harbored *S. sabaudiense* and 7.3% harbored the *Cantharis* spiroplasma, (25).

All isolates of msp have been obtained exclusively from female mosquitoes except for one exception, a strain of the *Cantharis* spiroplasma

TABLE 7.6. Mosquito species found infected by mollicutes in France.

	Savoia	Loire Atlantique
Spiroplasma sabaudiense	Ae. stiticus	Ae. detritus
Cantharis spiroplasma	Ae. cantans	Ae. detritus
	Ae. cinereus	Ae. caspius
	Ae. sticticus	
	Ae. rusticus	
	Coquillettidia richiardii	
Nonhelical mollicutes	Ae. cantans	Ae. detritus
	Ae. sticticus	Ae. caspius

from a male *Ae. caspius* caught in June 1988 in Loire Atlantique. Since the individual testing of mosquitoes began, all isolates have been obtained from whole-body homogenates and not from external washings. This clearly indicates an intracorporeal infection or symbiosis by msp. The actual location of the spiroplasma within the body of mosquito, that is, within the digestive tract or the hemolymph (or both) or eventually the salivary glands, remains to be clarified.

Ecological Factors and Msp Infection

During field studies conducted by our group in Savoia and Loire Atlantique, at least six ecological factors appeared potentially important: the species and the sex of the mosquito, the month and the method of mosquito collection, and the altitude of the biotope and the associated vegetation.

We have noted already that the species and the sex of the mosquito were important ecological factors for msp infection. All but one of our isolates came from females (90). We are unaware of the factor (or factors) that restrict msp infection to females. Concerning the importance of mosquito species, it is noteworthy that among other mosquitoes species tested in sufficient numbers *Aedes* sp., probably *Ae. cataphylla* ($n = 874$), in Jarrier (Savoia), and *Anopheles maculipennis* ($n = 424$), in Loire Atlantique were found negative for msp.

The importance of the month and the method of mosquito collection has been established by Chastel et al. (24, 25). From 1983 to 1988, repeated isolations of msp were achieved during June and July in Savoia and during May and June (1988) in Loire Atlantique. Subsequently, msp disappeared in August in Savoia and from August to September in Loire Atlantique. In the United States, *S. culicicola* has been isolated in August, whereas in Taiwan, *S. taiwanense* has been isolated in July.

Furthermore, all isolations of msp were acquired from mosquitoes caught from human or animal bait. In Savoia, no isolations have been made from mosquitoes caught from hood nets ($n = 765$), which excluded the possibility of flower contact following larval emergence, or from 500 larvae of *Ae. sticticus*. Accordingly, no isolations were obtained in Loire Atlantique from field-caught nymphs that emerged in the laboratory (90). These observations do not support the acquisition of msp infections by mosquitoes from their aquatic environment.

Finally, altitude of mosquito larval resting places and the associated vegetation may represent also important ecological factors. No isolation has been obtained from Jarrier at 1538 m (24) and this contrasted with multiple isolations of msp from biotopes of low (Isère Valley) or very low altitudes (Loire Atlantique). However, in Jarrier's mountainous biotopes, only *Ae. cataphylla* were caught in 1984 and 1985. Thus "altitude" and/or the "mosquito species" may have been responsible for these negative results.

Associated vegetation is a critical ecological factor for other spiroplasmas, such as bee- or beetle-derived spiroplasmas, and, therefore, might be relevant to msp ecology.

From 50 specimens of flowers, including 15 different species of plants collected in Savoia in July 1984, two strains of the *Cantharis* spiroplasma were isolated from the thistle, *Circium* spp. (25) (Table 7.7). However, 63 specimens of the same flowers. collected in July, August, and September 1987 from the same location produced no msp. It is noteworshy that *S. sabaudiense* has been never isolated from these flowers.

Population Dynamics of Mosquito Spiroplasmas

In the previous paragraph, we emphasized that the month of capture of mosquitoes appeared as a critical factor in msp ecology. This is also true for many insect-associated spiroplasmas. (63).

Accordingly, studies on population dynamics of *S. sabaudiense* and the *Cantharis* spiroplasma have been effected during the spring to summer months both in northern Alps (1987–1988) and in the Loire Atlantique district (1988). In the northern Alps, the survey concerned female mosquitoes belonging to eight different species that were caught from human bait, whereas in Loire Atlantique, both males and females belonging to nine different species were collected, and were caught either from human and animal bait or after having emerged in the laboratory. All specimens were examined individually.

In Table 7.8, the population dynamics of msp in Savoia has been constructed from data obtained in March and June 1988, and in July, August, and September 1987. The results of the survey carried out in Loire Atlantique on mosquitoes collected here during May, June, August, and September 1988 are presented in Table 7.9.

The results of the two surveys appeared to be very similar. In Savoia, no msp were detected until June; msp incidence increased rapidly in July and August, but became nil in September. In Loire Atlantique, msp appeared

TABLE 7.7. Isolations of Mollicutes from flowers, Savoia (July 1986).

Species	Number of tested specimens	Cantharis spiroplasma		Nonhelical mollicutes	
		W[a]	C[b]	W	C
Zea mays	17	—	—	—	—
Hyperricum perforatum	3	—	—	—	—
Spirea ulmaria[c]	6	—	—	1	3
Circium sp.[d]	5	2	—	—	—
Solidago serotina	3	—	—	—	—
Lotus corniculatus	2	—	—	—	—
Trifolium pratense	4	—	—	—	—
Lysimachia sp.	2	—	—	—	—
Convolvulus sepium	1	—	—	—	—
Lythrum salicaria	2	—	—	—	—
Aster	1	—	—	—	—
Achillea millefolium	1	—	—	—	—
Taraxacum officinale	1	—	—	—	—
Leucanthemum vulgare	1	—	—	—	—
Daucus carota	1	—	—	—	—
Total	50	2	0	1	3

[a] Washing.
[b] Crushing.
[c] Meadowsweet.
[d] Thistle.

TABLE 7.8. Dynamics of mosquito spiroplasmas isolations in Savoia (1987–1988).

	March 1988	June 1988 S. Sabaud	July 1987		August 1987		September 1987	Total
			S. Sabaud	Cantharis	S. Sabaud	Cantharis		
Ae. sticticus		0/15	5/80	1/80	1/122	4/122	0/11	11/228
Ae. cantans		1/49	0/45	0/45	0/12	0/12		2/106
Ae. vexans			0/12	0/12	0/14	0/14	0/12	0/38
Cu. annulata	0/36						0/1	0/37
An. maculipennis	0/22							0/22
Ae. geminus		0/5		0/20			0/2	0/7
Ae. rusticus			1/20					1/20
Ae. cinereus					0/8	0/8	0/1	0/9
Total	0/58	1/69	7/157	7/157	6/156	6/156	0/27	14/467

TABLE 7.9. Dynamics of mosquito spiroplasmas isolations in Loire Atlantique (1988).

		May 1988 (Sp. sabaudiense)		June 1988 (Cantharis spiroplasma)	August 1988 Catching	September 1988	Total
		Catching	After emerging				
Aedes detritus	♂	0/38[b]	0/5	—	—	—	0/43
	♀	1/37	0/23	8/205	0/57	0/19	9/341
Aedes caspius	♂	—	0/28	1/1	—	—	1/29
	♀+♀	—	0/13	1/53	0/95	0/41	1/202
An. maculipennis[a]	♀+♀	—	—	—	—	0/424	0/424
Miscellaneous[a]	♂	0/1	—	0/1	—	0/1	0/3
	♀	—	0/1	0/8	—	0/16	0/25
Total		1/76	0/70	10/268	0/152	0/501	11/1067

[a] Number of isolated spiroplasmas/number of tested mosquitoes.
[b] Including Anopheles claviger, Culiseta annulata, Cu.. subochrea, Cu. morsitans, Culex pipiens, and Aedes sp.

in May, were abundant in June, and disappeared during August and September (90; C. Chastel et al., unpublished).

The coincidence noted for msp incidence and the time of year may be explained easily, if msp were acquired by mosquitoes during spring from flowers as a result of nectar feeding. Nectar feeding by mosquitoes, both males and females, has been well documented in different countries (4, 6, 15, 57, 98, 99, 140, 141). Moreover, Giannotti and Giannotti, (54) have shown experimentally that a number of spiroplasmas can multiply in situ and in vitro in the flower nectar of various plants. Chastel et al. (25) recorded that most of their msp isolates grew readily in the nectar of *Kniphofia uvaria*, the natural well-balanced culture medium proposed by Giannotti and Giannotti (54).

However, it is necessary to remember that only the *Cantharis* spiroplasma has been found in flowers (*Circium* spp.) among 15 different species of plants tested during the 1986. This msp has not been found in 17 different plants during 1987. Surprisingly, the *Cantharis* spiroplasma was also the only one isolated from male mosquitoes, in spite of the fact that males readily feed on nectar. *S. sabaudiense* was never isolated from any flower.

During the 1987 to 1988 survey in Savoia, other insects such as leafhoppers, tabanid flies, and honey bees were tested for spiroplasmas. *S. melliferum*, a well-known bee pathogen described by Clark et al. in 1985 (32) was isolated on two occasions from a honey bee foraging on *Solidago serotina* and from a wild unidentified bee. The latter plant species is a preferred food source for *Aedes* mosquitoes.

Although *S. melliferum* circulated in nature together with two msp in Savoia, our group has not been able to isolate this spiroplasma from the large number of mosquitoes and flowers tested. In addition, msp were never isolated from insects other than mosquitoes. Thus the natural cycles of *S. sabaudiense*, the *Cantharis* spiroplasma, and *S. melliferum* appear well distinct, at least, in Savoia, although our knowledge of these cycles remains limited. Nonetheless, this finding may well indicate a certain degree of host specificity exhibited by msp for mosquitoes, however, the Sp. 34 strain, an isolate from *Ae. vexans* (Japan), related to *S. melliferum* (J.G. Tully, personal communication 1989) would tend to indicate otherwise.

Persistence

As for other Mollicutes, spiroplasmas were thought not to possess persistent forms or "resting stages" in their life history (79), even though they must survive in the natural environment and undergo transmission between plants, arthropods, and possibly vertebrates (16, 51).

However, recent observations would tend to indicate the contrary, at least for msp (71, 74, 76). Garnier et al. (51) reviewed the literature on this subject and presented evidence for *S. citri* that culture viability depended on the presence of helical forms. This would appear to be correct in arthropod-cell-free media (74). In C6/36 *Ae. albopictus* cell cultures infected with both *S. sabaudiense* and *S. taiwanense*, (SP2) cultures remained viable in the absence of helical forms and were characterized by the presence of numerous "ampullae" or persistent forms (Fig. 7.2f).

In the absence of medium change, growth and persistence at 28°C (Fig. 7.6) and persistence at 4 and 37°C (Fig. 7.7) were measured for *S. taiwanense* in cell-free spiroplasma growth medium (SP4) and in cell-associated culture (CAC). *S. taiwanense* persisted significantly longer ($p <$ 0.001) at all three temperatures in CAC (74). Although helical forms persisted longer in CAC, cultures remained viable long after the disappearance of helical forms. In addition, persistence in the face of total destruction of the cell sheet due to aging and an absence of medium change showed that persistence was independent of cell survival. Similar results were obtained also for *S. sabaudiense* (71, 76).

Persistence was found to be dependent upon "cell-contact," however, whether this persistence depended merely upon the presence of cells in the culture chamber or rather due to nonhelical intracellular stages giving rise to persistent "ampullae" remains to be determined. These findings may obviate the need for a host system during the winter months and associated periods of low ambient temperature. Furthermore, the death of the mosquito host may well be irrelevant to the persistence of msp. The dynamics of this persistence have yet to be clarified in the field.

Interestingly, *S. sabaudiense*, originally isolated from the French Alps, appeared to survive better at lower temperatures than did *S. taiwanense*, isolated from Taiwan ($p < 0.001$, t-test for comparisons of 2 slopes), whereas the reverse situation at warmer temperatures (37°C) showed *S. taiwanense* to survive comparatively longer. (NB Neither species persist for extended periods at 37°C.)

Mosquito Spiroplasmas and Vertebrate Hosts

Spiroplasmas are considered essentially as plant- or insect-pathogens or insect symbionts, and so far, there is no proof or indication that they might naturally infect vertebrates or induce any pathology in animals other than insects in nature.

However, *Spiroplasma mirum*, a tick-derived spiroplasma, previously known as "suckling mouse cataract agent," is able to multiply at 37°C and produce severe degenerative lesions of eyes and/or brain when inoculated intracranially into young mice, rats, syrian hamsters, and rabbits (30, 82, 83). Experimental *S. mirum* infection in suckling rats resembles the spongiform degenerative brain diseases ("spongiform encephalopathies") of both man and animals (12, 13). This has led to speculations about the eventual role of spiroplasmas in the etiology of Creutzfeldt-Jakob disease of man (see below).

As animals, including man, are bitten frequently by mosquitoes, the possibility that msp may infect vertebrates and may be responsible for

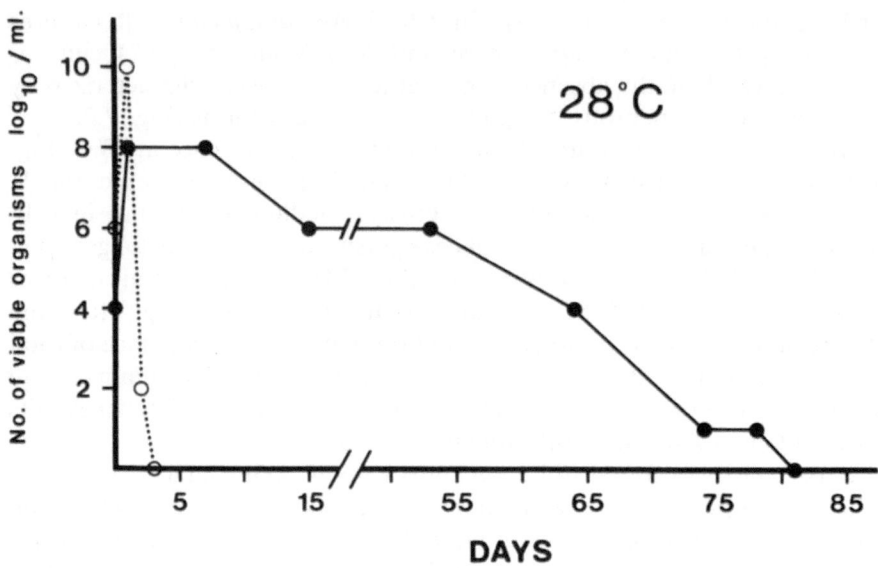

FIGURE 7.6. Growth and persistence of *Spiroplasma taiwanense* in SP4C (......)
and in CAC (———) at 28°C.

unsuspected pathological conditions should not a priori be neglected,
though, to date, no msp has been shown capable of replication at 37°C.

This stimulating aspect of msp biology led our group to undertake a
number of serosurveys in cattle, sentinel and wild animals, and healthy
blood donors from both Savoia and elsewhere. *S. apis*, *S. sabaudiense*, the
Cantharis spiroplasma, *S. taiwanense*, and in some instances Sp. 18 and
Sp. 34 strains were used as antigens in metabolic inhibition tests. When
positive results were observed they were verified using the deformation
test. Results of these investigations have been published already (24, 25).

A total of 565 healthy blood donors were examined, 499 of them living in
Savoia and the 66 others in Brittany. All these sera were found to be
negative. Two sentinel pigeons and two sentinel rabbits set out near Grésy
and Aiton (two localities from which msp have been isolated) were serially
bled from May to July 1985. Again, no antibodies to msp were detected.

Sera were also collected from 20 cows in Savoia, in November 1985,
November 1986, and December 1987. Low titers of antibody to the
Cantharis spiroplasma were found by both metabolic inhibition (1:4 to
1:16) and deformation tests (1:16 to 1:32) in four cows bled in November
1985. The same sera exhibited antibodies to a number of "bovine"
mycoplasmas but there was no clear relationship between these later
antibodies and "*Cantharis*" antibody, thus excluding antibody cross-
reactivity.

Some 366 sera from wild, small mammals (rodents, insectivores) were

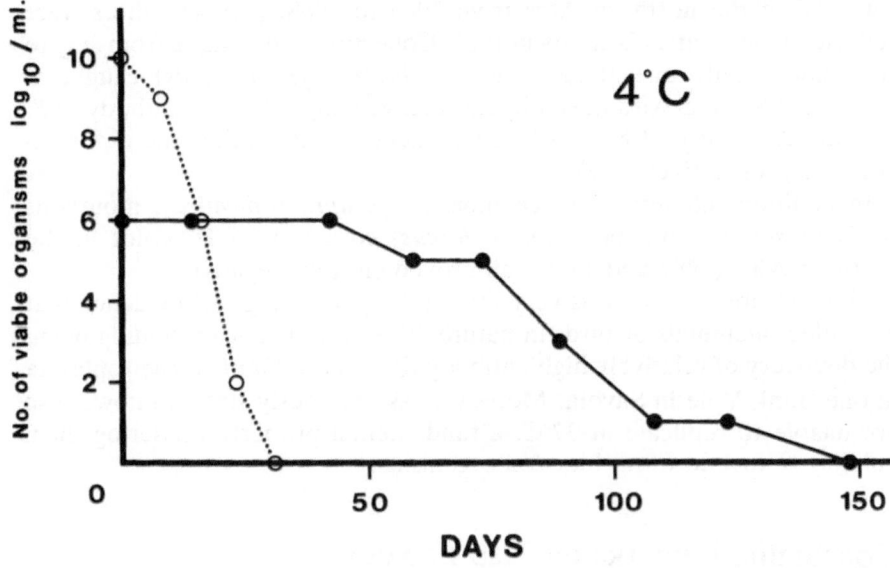

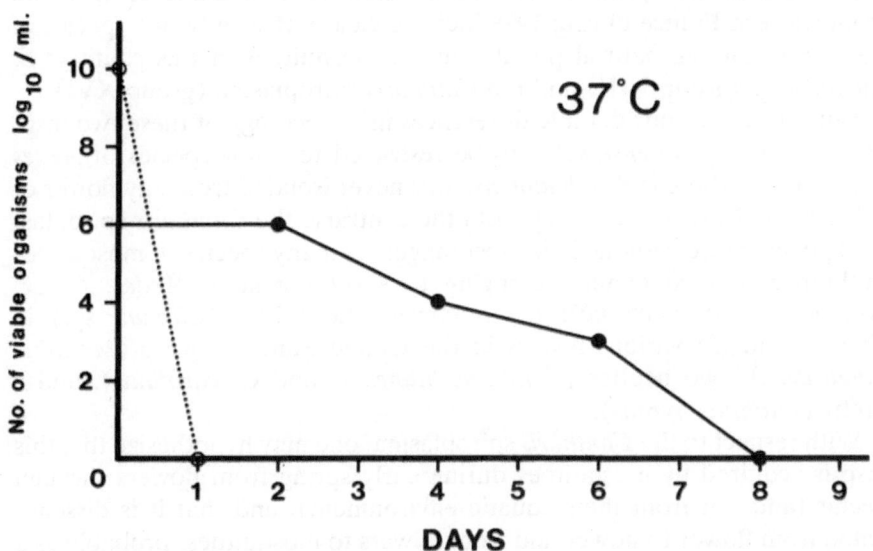

FIGURE 7.7. Persistence of *Spiroplasma taiwanense* in SP4C (......) and in CAC (_____) at 4 and at 37°C.

collected in the northern Alps from 1985 to 1988, plus 41 others were collected from central Brittany in 1987. Collections were made from July to November. Only one Bank Vole (*Clethrionomys glareolus*) caught in October 1986 near Aiton exhibited a relatively high titer of antibody to *S. sabaudiense*, that is, 1:80 and 1:160 by metabolic inhibition and deformation tests, respectively (25).

In addition, 68 sera from chamois (*Rupicapra rupicapra*), moufflons (*Ovis musimon*) and ibex (*Capra hircus*) collected by B. Gilot in the northern Alps (1987 and 1988), all proved entirely negative.

Thus, at present, there is very little indirect (serological) evidence that msp infect mammals or birds in nature. The only significant finding being the discovery of relatively high antibody titer to the *Cantharis* spiroplasma in one Bank Vole in Savoia. Moreover, as previously stated, known msp are unable to replicate at 37°C, a fundamental property rendering them unlikely to invade warm-blooded organisms.

Concluding Remarks on Msp Ecology

The results of ecological studies accumulated from Savoia since 1983 and from western France during 1988 indicate clearly that only two species of msp occur among natural populations of mosquitoes in this country: *S. sabaudiense* (group XIII) and the *Cantharis* spiroplasma (group XVI).

But, there are considerable differences in the ecology of these two msp. Hosts of *S. sabaudiense* seem to be restricted to a few species of *Aedes* mosquitoes, whereas *S. sabaudiense* was never isolated from any flower or other insect during our surveys. On the contrary, the *Cantharis* spiroplasma appears more catholic in its host range: (1) many species of mosquitoes in France and Alabama, belonging to several genera (*Aedes*, *Culex*, *Anopheles*, *Coquillettidia*); (2) a flower, the thistle (*Circium* sp.) in France; and (3) various insects in the United States: a wasp (*Monobia quadridens*), two beetles (*Cantharis bilineatus* and *C. carolinus*), and a firefly (*Photinus pyralis*).

With respect to the *Cantharis* spiroplasma, one may hypothesize that this msp is acquired by mosquitoes during early spring from flowers through nectar (and not from their aquatic environment), and that it is disseminated from flower to flower and from flowers to mosquitoes, probably as a consequence of a continuous contamination by other insects. It then disappears in the early fall as a consequence of some critical change in biotopes concerning flowering plants. At present, no explanation can be proposed to explain both the absence of infection in males and the life history of *S. sabaudiense*.

Yet there remains many unanswered questions. The more fundamental of these is what happens to msp during winter and how and where does overwintering of msp take place? Does it occur in plants, in hibernating mosquitoes, or other insects, in other invertebrates, or in cold-blooded vertebrates? Are the recently described (74) persistent forms of msp observed in vitro sufficiently resistant to adverse conditions to ensure the survival of the species in nature? Further investigations are necessary in order to understand this mystery. These findings are also likely to be applicable to other flower- and insect-associated mollicutes!

Pathogenicity

Spiroplasmas are known to be pathogenic for a wide variety of insect species both in vivo and in vitro (37, 70, 97).

Pathogenicity In Vivo

Using a laboratory isolate of *S. taiwanense* (SP2/mut/sl/P9) originally selected for its ability to lyse *Ae. albopictus* C6/36 cells in vitro, we were able to demonstrate a significant reduction in the survival of adult males and females of *Ae. aegypti* and *An. stephensi* following intrathoracic inoculation (73). These results are presented in Figure 7.8. The low level of mortality observed 24-h postinoculation (p.i.) for these two important vector species (0.25% and 4.8%, respectively with $n = 400$ for both species) would suggest that intrathoracic inoculation was associated with little trauma in adult male and female populations. Using Chi-squared tests, it was shown that survival had been significantly ($p < 0.001$) reduced at the 50, 90, and 95% mortality level for the test groups, except the 50% mortality level in adult male *An. stephensi*.

More importantly for vector species, significant flight reduction in infected adult female populations was demonstrated as of days 5 to 8 p.i. (Fig. 7.9) and day 4 p.i. (Table 7.10) for *Ae. aegypti* and *An. stephensi*, respectively. This flight reduction occurred prior to the earliest known extrinsic incubation period for Dengue virus in *Ae. aegypti*, that is, 11 to 14 days (118, 125) and that of malaria, 7 or more days in *Anopheles* mosquitoes (100). Thus in both cases, disease transmission would have been rendered highly improbable.

Three dilutions of *S. sabaudiense* introduced intrathoracically into adult female populations of *Ae. aegypti* failed to produce any significant reduction in survival when compared to equivalent control populations [I. Humphery-Smith et al., unpublished).

The larvicidal action of *S. taiwanense* (SP2 and SP2/mut/sl/P9) has been demonstrated, whereby per os exposure with helical replicative forms significantly reduced survival of *Ae. aegypti* (75). These results were obtained using 23 paired tubes (controls and test group) containing 25 first-stage larvae in 10 ml of distilled water for *S. taiwanense* (SP2) and using 10 paired tubes for *S. taiwanense* (SP2/mut/sl/P9) (Tables 11 and 12).

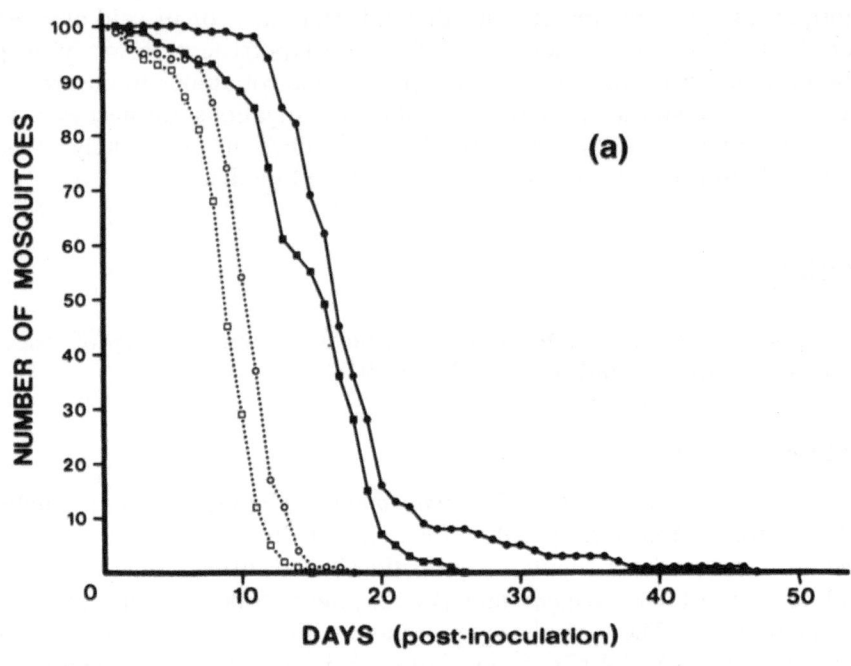

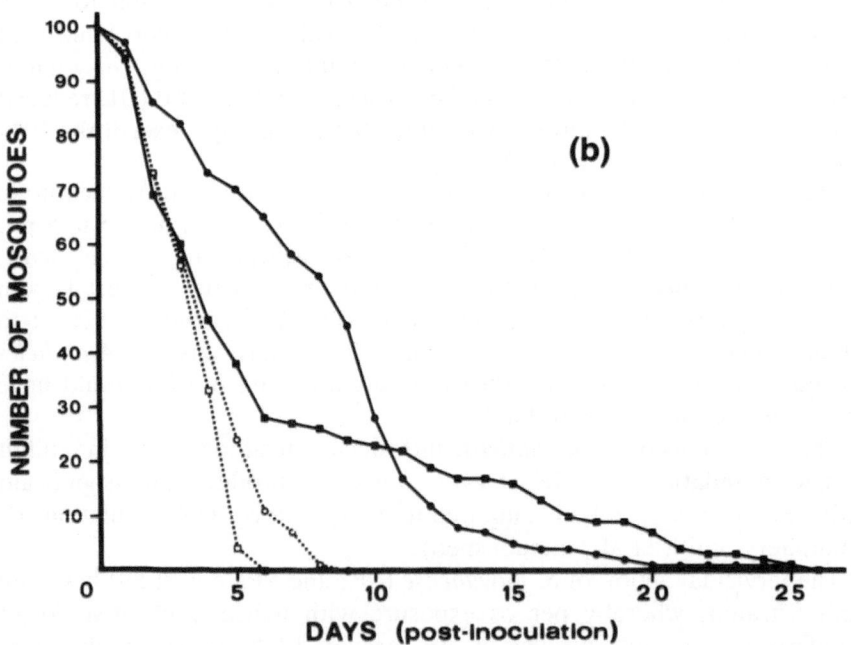

FIGURE 7.8. Reduced survival of a, *Aedes aegypti* and b, *Anopheles stephensi* following intrathoracic inoculation of *Spiroplasma taiwanense* (SP2/sl/mut/P9). Test group, ...; controls, ———; circles, females; squares, males.

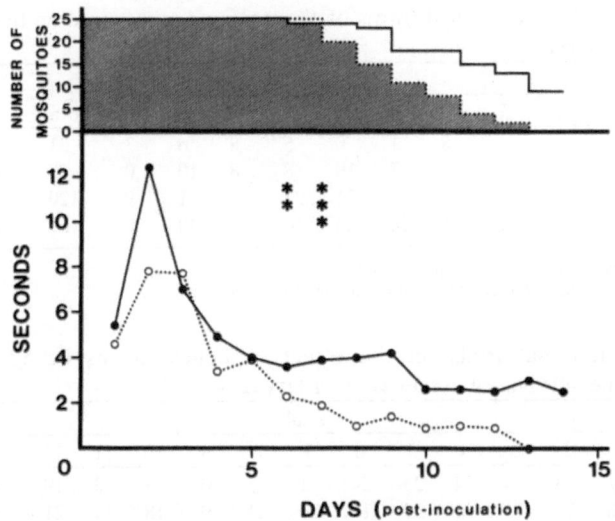

FIGURE 7.9. Reduced flight capacity and reduced survival of adult female *Aedes aegypti* induced by *Spiroplasma taiwanense* (SP2/sl/mut/P9) following intrathoracic inoculation. Flight reduction was measured by "Drop Tests" and each point represents the mean of 10 replications. Test group, ...; controls, _____; ** $p < 0.02$; *** $p < 0.01$ (Wilcoxon Sign Ranked Test).

TABLE 7.10. The effect of *Spiroplasma taiwanense* on the flight capacity of *Anopheles stephensi* at 4 days postinoculation.

Trial	Control group ($n = 22$)	Test group ($n = 29$)
1	6.3[a]	1.3
2	5.0	1.4
3	5.8	3.6
4	6.7	2.2
5	4.5	2.2
6	3.4	2.3
7	5.4	2.2
8	5.0	1.9
9	7.0	2.6
10	6.5	2.4
11	8.6	1.4
12	5.0	1.9
13	6.1	1.9
14	5.8	1.8
15	10.7	1.6
Mean	6.1	2.1

[a] Flight time in seconds.

TABLE 7.11. The effect of helical forms of *Spiroplasma taiwanense* on the survival of *Aedes aegypti* larvae.

Strain	Trial										Total[b]	value
	1	2	3	4	5	6	7	8	9	10		
Sp-2[a]	0	0	0	8	1	0	5	8	6	3	31	p < 0.001
Controls	0	16	11	3	7	10	8	8	10	0	73	p < 0.001
Sp-2/sl/mut/P9	21	6	10	1	7	21	22	22	1	9	120	p < 0.001
Controls	16	19	6	21	24	23	16	12	20	20	167	p < 0.001

[a] Total number of nymphs obtained from 25 larvae aged < 24 h.
[b] Total number of nymphs obtained from 250 larvae aged < 24 h.

TABLE 7.12. Additional trials on the effect of helical forms of *Spiroplasma taiwanense* on the survival of *Aedes aegypti* larvae.

S. taiwanense	Trial													Total[b]
	1	2	3	4	5	6	7	8	9	10	11	12	13	
Sp-2[a]	7	0	0	24	23	24	1	3	0	3	3	19	3	110
Controls	16	21	22	21	22	17	23	24	9	18	24	21	15	253

[a] Number of nymphs obtained from 25 larvae aged < 24 h.
[b] Number of nymphs obtained from 325 larvae aged < 24 h.

Larvae were maintained on a diet consisting of a minimal quantity of yeast administered every 2 days. This protocol using first-stage larvae was adopted to avoid the variability noted in eclosion rates, even when eggs were taken from the same oviposition, that is, the same piece of filter paper from the same cage of adult *Ae. aegypti*.

Disappointingly, the persistent forms of these two strains of *S. taiwanense* showed no pathogenicity for larval *Ae. aegypti* (75), whereas the helical forms of *S. taiwanense* possessed very limited survival capacity, that is, less than 72 h at 28°C. At present, it is not known under what circumstances in nature persistent forms give rise to replicative forms. In the laboratory, any medium sufficiently rich in nutriments is sufficient to provoke replication of these forms and presumably the capacity for pathogenicity. Simitzis et al. (126) have also demonstrated the msp SP7 isolated from *Armigeres subalbatus* in Taiwan to be pathogenic towards eggs of *Ae. aegypti*.

As yet, our investigations have not quantified spiroplasma replication in mosquito hosts, nor identified the predeliction sites. As we now dispose of a medium specific for msp, it will be possible to conduct reisolations of msp from aquatic media, infected mosquitoes, and varied locations without fear of losing nonhelical and cell-associated forms due to the filtration of tissue homogenates prior to passage in spiroplasma growth medium.

Pathogenicity In Vitro

Humphery-Smith et al. (70) described cytopathogenicity in C6/36 *Ae. albopictus* associated with infection of *S. sabaudiense*. Although the results observed were replicated three times in parallel, an intensive effort to increase pathogenicity by rapid passage failed to detect that the pathogenicity observed was due to contamination by another msp, *S. taiwanense*. The total destruction of cell sheets (Fig. 7.10) meant that immunofluorescent tests were no longer practicable and deformation tests were not conducted between passages. Since then efforts to reproduce the same level of cytopathogenicity using *S. sabaudiense* and *S. taiwanense* (SP2) in rapid passage (up to nine passages) have failed, but *S. taiwanense* was observed to retard cell growth significantly. Following further passage, the laboratory isolate responsible for the above pathogenicity, *S. taiwanense* (SP2/mut/sl/ P9), was found to be capable of extensive cell lysis and of provoking culture death in less than 40 h postinoculation, as determined by an absence of the cell sheet.

Increased passage in C6/36 cells was associated with increased pathogenicity, however, prior to the onset of cell lysis in less pathogenic strains, cytopathogenicity included the production of syncytia, an increase in the number of mitochondria and an alteration in their morphology, vacuolization, and a reduction in the rate of cell growth. Similar cytopathology and culture death has been recorded in vitro also for a number of other spiroplasmas (93, 94, 101, 131, 132).

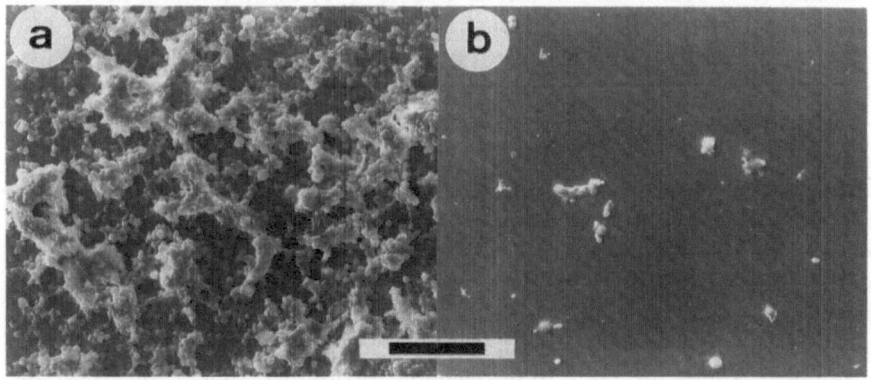

FIGURE 7.10. *Aedes albopictus* C6/36 cell cultures 6 days following passage. a, Uninfected and b, Infected with *Spiroplasma taiwanense* (SP2/sl/mut/P9). Scale bar, 100 µm.

Creutzfeldt-Jakob Disease (CJD) and Spiroplasmas

Creutzfeldt-Jakob Disease (CJD) and the closely related Gerstmann-Sträussler Syndrome are rare central nervous system disorders of man caused by an infectious agent for which the fundamental nature remains uncertain (42, 49). The filtrable agent responsible for these diseases in man is thought to be related to scrapie in sheep, kuru in residents of Papua-New

Guinea, transmissible mink encephalopathy, transmissible spongiform encephalopathy of mule deer and elk, and the recently discovered (November 1986) bovine spongiform encephalopathy (5, 49, 68, 102, 143). Understanding the infectious processes responsible for these slowly developing diseases (approximately 30 years in the case of kuru) will no doubt prove to be as great a breakthough as were the original demonstrations of viruses and bacteria as causative agents of disease.

At present, at least two schools of thought have been proposed to explain the occurrence of these diseases. The most commonly cited is that of subviral "prions" (108), which are thought not to possess a nonhost antigenic subunit protein (48) and, prior to the work of Narang et al. (104), not to be associated with nucleic acids. The unconvential nature of these infectious particles, the occurrence of prion protein (PrP) in healthy tissues, and the difficulty of other researchers (109) to attribute infectivity to proteinase-K-resistant PrP have cast doubts on the work implicating PrP.

Another school of thought implicates spiroplasmas. Some of these organisms (*S. mirum*) have been shown to be neurotropic and to produce spongiform encephalopathy experimentally in rats (12, 13). Bastian (10, 11), Reyes and Hoening (115), and Gray et al. (55) have reported spiroplasma-like inclusions in individuals suffering from CJD. The latter was shown by Humphery-Smith & Chastel (69) to be nothing more than crystalline artifacts, however, Bastian et al. (14) have been able to show that fibril protein purified from *S. mirum* and scrapie-associated fibril protein are related serologically using Western Blots.

The thermophilic nature of *S. mirum*, viz, its ability to grow and persist at 37°C, is the feature that sets it apart from other known spiroplasmas. The pathology associated with this species has been reviewed recently (97). The msp described to date have not proved to be thermophilic. However, our knowledge of msp and other spiroplasmas transmitted by blood-feeding arthropods has only just started to be realized. As an example, from just 16 homogenates, we were able to obtain recently 9 isolates of spiroplasmas from 7 different species of Tabanidae in France [I. Humphery-Smith et al., unpublished). Prior to these observations, spiroplasmas had not been described previously from tabanids in Europe. As more hematophagous arthropods are examined for the presence of spiroplasmas, it is quite plausible that other thermophilic species will be found. Given the possible medical and veterinary importance of spiroplasma transmission by hematophagous arthropods, it is hoped that further investigations will be initiated in the near future in other parts of the world.

Using another thermophilic species, *S. melliferum*, Chastel et al. (23) have been able to show neurological symptoms associated with intracerebral inoculations of suckling mice and persistence in vivo for up to 9 months. As with some msp, *S. melliferum* is thought to be acquired from flowers by honey bees (63).

Similarities shown by spiroplasmas and the infectious material known to give rise to spongiform encephalopathies include: filtrability (80, 127); slow persistent infection associated with neurotrophy (13); resistance to a wide range of fixatives such as alcohol, phenol, glutaraldehyde, formalin (130), sodium hypochlorite (cf. Appendix 7.2); resistance to some heavy-metal salts (147), and resistance to a wide range of antibiotics (1, 17, 20, 40, 84, 92, 123). Notable among the disimilarities between spongiform encephalopathies and spiroplasma infections are susceptibility to ultraviolet radiation (123, 87) and production of antibody during infection. (NB This is not so following intracerebral inoculation).

Strangely, homologous antibody has been shown to prolong the viability of *S. sabaudiense* in cell-free growth medium (76). Presumably, the antibody or aggregations of spiroplasmas provide a protective proteinaceous coat surrounding viable spiroplasmas, not unlike preparations of PrP. Garnier et al. (50–53) and Bové (16) have demonstrated that spiroplasmas have the ability to reproduce both extra- and intracellularly, whereas their reduced genome size and the absence of a cell wall gives these prokaryotes capabilities of pleomorphism and many physiological characters not found in other free-living organisms. In addition, using prolonged centrifugation, Humphery-Smith et al. (76) showed that spiroplasmas swell up to form "balloons" too large to pass through filters (0.2 μm) and yet filtrable debris remained capable of reproducing helical morphology.

As a direct approach towards clarifying the etiological role of spiroplasmas in CJD, microbiological and serological studies have been carried out in patients with CJD or other degenerative central nervous system (CNS) disorders.

Leach et al. (89) tried to detect spiroplasmas by both cultivation in SP4 medium and by serological tests on 18 patients. No spiroplasmas or other mycoplasmas were cultivated from brain tissue of 18 cases and no antibody was demonstrated to several spiroplasmas belonging to Groups I (*S. citri*, *S. melliferum*, 277 F), III (*S. floricola*), IV (*SR-3*) and V (*S. mirum*). These negative results were confirmed by Chastel et al. (26) who tested 255 sera from patients and healthy individuals against five msp isolated in France, Taiwan, and Japan. Furthermore, no antibody was demonstrated in 35 patients with multiple sclerosis, in 3 with CJD and in 1 with Alzheimer's sclerosis. However, low antibody titer to *S. sabaudiense* was found in a 35-year-old patient with amyotrophic lateral sclerosis. Given the extreme rarity of these conditions and thus the limited sample sizes, it is difficult to draw meaningful conclusions from such studies.

Mechanisms of Msp Pathogenicity

Cytadsorption

Humphery-Smith et al. (70) observed synchrony in the appearance of cell lysis and cytoadsorption of *S. taiwanense* in their cultures of C6/36 *Ae.*

albopictus cells. This interpretation was consistent with scanning electron micrographs showing cytadsorption resulting in an increase in cell size presumably followed by membrane lysis (Fig. 7.11). Eskafi et al. (41) reported vast destruction of cells and loss of membrane integrity in *Galleria mellonella* infected with *S. floricola*, which they concluded may have been due to attachment of spiroplasmas to host cells. Elsewhere, Steiner et al. (131, 132) demonstrated cell culture death caused by spiroplasmas and suggested that cytadsorption and cytopathogenicity were probably linked. Again this is thought to be the case for mycoplasma infections (111), yet has been difficult to quantify. For this reason, the overt cell lysis observed in vitro for *S. taiwanense* may well provide a useful model for understanding the mechanisms of pathogenicity among mollicutes, important disease-causing organisms of medical and veterinary importance. The widespread presence of mycoplasmas in cell cultures in laboratories around the world is evidence of the more subtle pathogenic processes that take place in mycoplasma infections. Such infections can go unnoticed for extended periods.

Our initial efforts to quantify cytadsorption using nonpathogenic *S. sabaudiense* and pathogenic *S. taiwanense* (SP2/mut/sl/P9) radioactively labelled with ^{14}C-marked sphingomyelin, a phospholipid for which spiroplasmas show an avidity (62), showed no difference in the rate of cytadsorption to C6/36 *Ae. albopictus* cells [I. Humphery-Smith et al., unpublished). These observations do not preclude cytadsorption from being associated with, another event, which for example, deregulates host-cell membrane permeability in pathogenic strains.

Biological Activity

Spiroplasmas are capable of very rapid growth and reproduction to attain high concentrations (10^{12} CCU/ml in < 48 h) in nutrient-rich media. This biological activity in vitro is associated with an acidification of the culture medium, for example, pH 7.5 to pH 6.6 and presumably a reduction in the available nutrients in the growth medium. Again, similar reductions in the pH of culture media have been observed in both pathogenic and non-pathogenic strains and species. Eskafi et al. (41) produced histopathological observations on *Galleria mellonella* infected with *S. floricola*, which showed low or no gut contents, breakdown of gut epithelium, a drop in hemolymph glycogen, disintegration of connective tissue, and consumption of or breakdown of fat body. As mentioned earlier, spiroplasmas ferment carbohydrate substrates to generate ATP by glycolysis. This substrate in insects is composed primarily of glycogen, which occurs in many of the insect organs where Eskafi et al. found spiroplasmas. Their sterol requirements in insects are furnished by cholesterol (cf. 41), and *S. floricola* was found in close association with cholesterol-bearing structures such as subcellular membranes, hemolymph, fat bodies, and blood cells. It

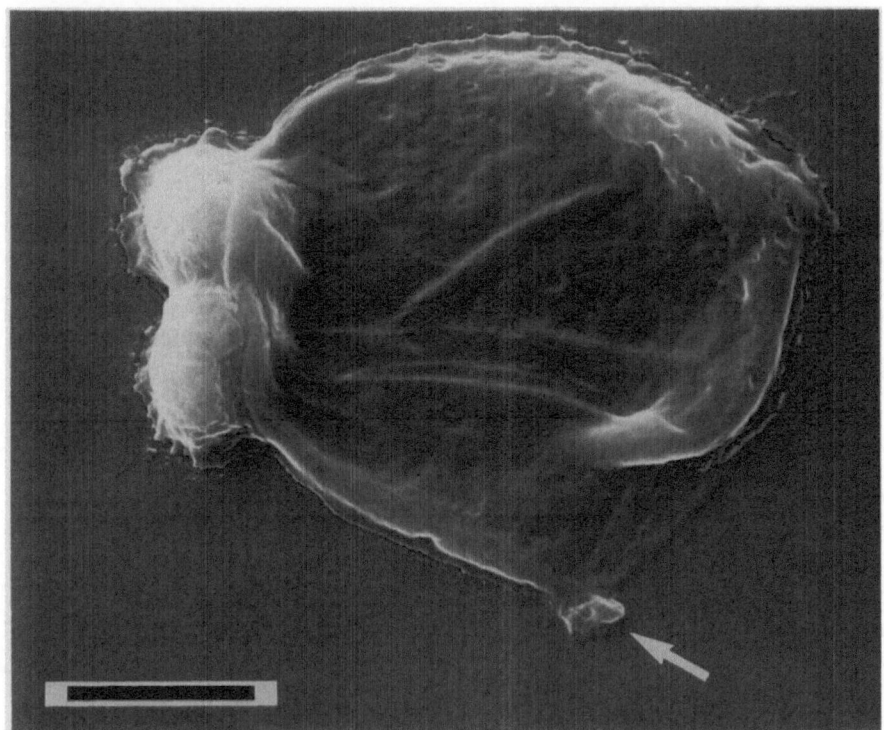

FIGURE 7.11. Cytadsorption of an *Aedes albopictus* cell (C6/36) in vitro by *Spiropalasma taiwanense* (SP2/sl/mut/P9) intermediate between a helical and a coccoid form (arrowed). Two uninfected cells appear to the left of the infected swollen cell. Scale bar, 10 μm.

is evident that the energetic demands placed on an infected organism to support up to 10^{10} *S. floricola* in the above instance, may well explain spiroplasmas pathogenicity.

Toxins

The existence of spiroplasma toxins has been proposed by Daniels (35–37), whereas recent observations on spiroplasma phylogeny would tend to indicate that these organisms were derived from (toxin-producing) *Clostridium*-like ancestors (154; and J.M. Bové, personal communication, 1989).

Plasmids

In experimental infections with *Salmonella typhimurium*, the presence of certain plasmids has been correlated with several virulence factors including the ability to penetrate into epithelial cells, to resist the bactericidal action of normal serum, and to multiply in the liver and spleen (67, 81, 105).

Although plasmids have been described in a wide variety of spiroplasmas, they have yet to be described in msp and, at present, are all considered to be cryptic (16, 97).

Spiroplasma viruses

Some four spiroplasma-associated viruses have been reported in *S. citri* (16, 97), however they do not appear to play a role in the production of disease symptoms. Rather, they may well reduce pathogenicity by lysing spiroplasmas. Again, these viruses have not been shown in msp yet.

Alteration of Male-Female Sex Ratio

Another spiroplasma, the Sex Ratio Organism, is known to be capable of biasing the sex ratio of progeny of infected *Drosphila* fruit flies in favor of females (150). In an attempt to use this trait as an eventual means of controlling mosquito vectors, Fitz-Earle et al. (43) infected female pupae and adults of *Toxorhynchites splendens*. Unfortunately, sex ratio distortion was not achieved. These experiments, however, were able to demonstrate transovarial transmission of this organism in *T. splendens*.

Transformation

Transformation of NIH3T3 cells in the presence of *S. mirum* has been demonstrated by Kotani et al. (85). The continued presence of *S. mirum* was shown not to be necessary for the maintenance of the transformed phenotype. As carried out in a cell line capable of spontaneous transformation (56), this work cannot be regarded as conclusive, but does offer yet another potential mode of pathogenicity open to msp. However, these results further stress the need to handle msp and other spiroplasmas with caution in the laboratory.

Potential for the Use of Msp in Mosquito Vector Control

Attempts to control mosquito-borne diseases have failed largely because of (1) the diversity of larval breeding sites of vectors; (2) manpower and material costs associated with administering insecticides and surveillance of breeding sites, (3) vector resistance to insecticides; (4) drug resistance of infectious agents; (5) low-target specificity and ecological nondesirability of insecticides; and (6) the difficulty of developing efficient polyvalent vaccines. For these and other reasons, Halstead (65), Reeves (114), and Gubler (59) have called for novel approaches to mosquito vector control, and in particular, with respect to *Aedes* mosquitoes. With this optic, we have been examining mosquito spiroplasmas as potential biological control agents of mosquito vectors (*Aedes, Anopheles*, and *Culex*). The major tropical diseases they transmit (Malaria, Filariasis, Dengue Fever, Yellow Fever) remain an imposing challenge to world health authorities.

Alone, *Aedes aegypti*, the principal Dengue vector, has a pantropical distribution delimited by the midwinter 10°C isotherm (a region home to almost half the world's population), whereas malaria kills more people per year than AIDS will have killed before the year 2000. In addition, mosquitoes are of considerable nuisance value when found in association with many of the world's major cities both in developed and third world countries.

S. taiwanense would appear to offer much promise in this area. It has been shown to (1) significantly reduce the flight capacity of females and the survival of adult males and females of *Ae. aegypti* and *An. stephensi* following intrathoracic inoculation; (2) significantly reduce survival following per os infection of larval *Ae. aegypti*; (3) possess a persistent form in its life history; (4) lyse *Ae. albopictus* C6/36 cells in vitro; and most importantly, (5) have a reproductive capacity typical of prokaryotes. Furthermore, *S. taiwanense* is considered an unlikely pathogen of warm-blooded vertebrates due to its inability to persist at 37°C in vitro and the absence of observable pathogenicity and reisolation following intracerebral inoculation of day-old suckling mice. (74; C. Chastel, unpublished). The latter is a property shared by *S. sabaudiense* and the *Cantharis* spiroplasma (69; C. Chastel, unpublished).

Larval infectivity offers a means whereby this bacterium might be introduced into wild mosquito populations, however, our immediaty priority is to identify the molecular basis of this pathogenicity, as (1) it would be unwise to undertake large-scale release of an infectious agent in the field without this knowledge, and (2) such understanding may well allow the gene(s) responsible for this pathogenicity to be cloned and expressed commercially.

Another alternative is being considered also in our laboratory, whereby a nonpathogenic msp, such as *S. sabaudiense*, might feasibly be used to infect mosquito populations and similtaneously express a deleterious factor previously cloned into this organism. The presumed host specificity of *S. sabaudiense* would be a further advantage associated with such a strategy. Furthermore, we still lack details concerning the whereabouts of msp in their mosquito hosts and their survival and reproduction in mosquitoes. Nonetheless, before any such program could be initiated, the safety guidelines set down by the World Health Organisation, Special Programme for Research and Training in Tropical Diseases would necessarily have to be fulfilled.

Other bacterial agents known to be pathogenic towards mosquitoes are limited to the genus *Bacillus* (19). *B. thuringiensis* is effective against a large range of insect vectors, including members of *Aedes*, *Anopheles*, *Culex*, and *Simulium*. Although possessing a spore, *B. thuringiensis* does not persist for extended periods in larval breeding grounds, it sediments in liquids, and does not recycle within mosquito populations. Thus repeated applications of this biological control agent are required. These problems are not shared by *B. sphaericus*, however, this species is poorly pathogenic towards *Aedes* mosquitoes. Although a variety of organisms have been used or considered as potential biological control agents for mosquito vectors, it is perhaps bacteria and other prokaryotes that offer the greatest hope for success in this area because of their high reproductive capacity.

Examples of vector eradication being associated with the disappearance of disease (Yellow fever and/or Dengue fever) can be found in Singapore, Cuba, and Brazil (64). Unfortunately, because of vector reappearance, both Cuba and Brazil have suffered Dengue outbreaks subsequently (86, 122). Thus some hope is justified for an eventual erradication or elimination of vector-transmitted disease by reduction of vector populations, whereas any pressure that can be exerted on vector populations as part of integrated control strategies is desirable.

Summary and Conclusion

Mosquito spiroplasmas (msp) although recently discovered (1982) have rapidly created considerable interest among entomologists and microbiologists. They exhibit the same general properties as those of other spiroplasmas, and, therefore, may be studied by the same morphological, biological, and molecular methods.

Only three species are fully described presently, *S. culicicola* from the United States, *S. sabaudiense* from France, and *S. taiwanense* from the Far East. Another species, the *Cantharis* spiroplasma, though isolated from mosquitoes has been recovered also from other insects such as beetles, a wasp, a firefly, and a plant (*Circium* sp.). It represents possibly a cluster of closely related serovars or species.

Since 1985, ecology of msp has been studied by our group from different areas in France but only partial conclusions may be drawn at present. The *Cantharis* spiroplasma seems to be acquired from flowers through nectar feeding in early spring and then colonizes mosquito populations during the summer months only to disappear rapidly in autumn. Inexplicably, female mosquitoes are infected, but not males. A large number of mosquito species (*Aedes*, *Anopheles*, *Culex*, *Copriluttidia*) were found infected in both France and the United States. On the contrary, very little is known regarding the life history of *S. sabaudiense* or other msp.

One important question about msp concerns their potential pathogenicity for mosquitoes, other invertebrates, and warm-blooded vertebrates including an eventual role in Creutzfeldt-Jakob disease. Our group has established good experimental evidence that *S. taiwanense* significantly reduces the life span of certain mosquitoes vectors, such as *Aedes aegypti* or *Anopheles stephensi*. To date, msp have been shown to be incapable of multiplication at 37°C and there is no epidemiological proof that they might infect man or other mammals. This clearly makes msp good candidates for use in biological control against mosquito vectors of human and animal diseases.

Obviously, more studies are needed on these enigmatic organisms especially in the fields of ecology and pathogenicity mechanisms, however, data presently accumulated has delineated many promising avenues for future investigations.

Acknowledgments. This work was supported in part by grants from the "Intitut National de la Santé et de la Recherche Médicale" (CRE 85 80 06), "Direction des Etudes et des Recherches Techniques" (Subventions 86-067 and 89-230), Paris, and "Fondation Langlois", Rennes. The authors are greatly indebted also to B. Gilot, B. Devau, and R. Gruffaz, Grenoble, and M. Marjolet and J. Guilloteau, Nantes, for kindly providing us with ecological data on mosquitoes. Thanks are also devoted to all members of our Department, and more precisely to F. Le Goff, O. Grulet, A.M. Simitzis-Le Flohic, G. Le Lay, M.L. Abalain-Colloc, G. Kerdraon, M. Mellouet, and M. Odermatt.

References

1. Abalain-Colloc, M.L., Le Goff, F., Abalain, J.H., and Chastel, C., 1986, Sensibilité à divers antibiotiques de six souches de spiroplasmes isolées de moustiques en France, *Pathol. Biol.* **34**:360–363.
2. Abalain-Colloc, M.L., Chastel, C., Tully, J.G., Bové, J.M., Whitcomb, R.F., Gilot, B., and Williamson, D.L., 1987, *Spiroplasma sabaudiense* sp. nov. from mosquitoes collected in France, *Int. J. Syst. Bacteriol.* **37**:260–265.
3. Abalain-Colloc, M.L., Rosen, L., Tully, J.G., Bové, J.M., Chastel, C., and Williamson, D.L., 1988, *Spiroplasma taiwanense* sp. nov. from *Culex tritaeniorhynchus* mosquitoes collected in Taiwan, *Int. J. Syst. Bacteriol.* **38**:103–107.
4. Andersson, I.H. and Jaenson, T.G.T., 1987, Nectar feeding by mosquitoes in Sweden, with special reference to *Culex pipiens* and *Cx torentium*, *Med. Vet. Entomol.* **1**:59–64.
5. Anon., 1988, BSE and scrapie: agents for change, *Lancet* **ii**:607–608.
6. Ardö, P., 1958, On the feeding habits of the scandinavian mosquitoes, *Opus. Entomol.* **23**:171–191.
7. Barile, M.F. and Grabowski, M.W., 1978, Mycoplasma-cell culture-virus interactions: A brief review, Mc Garrity, G.J., Murphy, D.G. and Nichols, W.W., (eds): in Mycoplasma Infection of Cell Cultures, Plenum Press, New York, pp. 135–150.
8. Barile, M.F., 1979, Mycoplasma-tissue cell interactions, Tully, J.G. and

Whitcomb, R.F., (eds): in The Mycoplasmas, Vol. II, Academic Press, New York, pp. 425–474.

9. Bastardo, J.W., Ou, O.D., and Bussell, R.H., 1974, Biological and physical properties of the Suckling Mouse Cataract Agent grown in chick embryos, *Infect. Immun.*, **9**:444–451.

10. Bastian, F.O., 1979, Spiroplasma-like inclusions in Creutzfeldt-Jakob disease, *Arch. Pathol. Lab. Med.* **103**:665–669.

11. Bastian, F.O., 1981, Additional evidence of spiroplasma in Creutzfeldt-Jakob disease, *Lancet*, **i**:660.

12. Bastian, F.O., Purnell, D.M., and Tully, J.G., 1984, Neuropathology of spiroplasma infection in the rat brain, *Am. J. Pathol* **114**:496–514.

13. Bastian, F.O., Jennings, R.A., and Hoff, C.J., 1987, Neurotropic response of *Spiroplasma mirum* following peripheral inoculation in the rat, *Ann. Inst. Pasteur Microbiol.* **138**:651–655.

14. Bastian, F.O., Jennings, R.A., and Gardner, W.A., 1987, Antiserum to scrapie-associated fibril protein cross-reacts with *Spiroplasma mirum* fibril protein, *J. Clin. Microbiol.* **25**:2430–2431.

15. Biblingmayer, W.L. and Hem, D.G., 1973, Sugar feeding by Florida mosquitoes, *Mosq. News* **33**:535–538.

16. Bové, J.M., Carle, P., Garnier, M., Laigert, F., Renaudin, J., and Saillard, C., 1989, Molecular and cellular biology of spiroplasmas, Whitcomb, R.F. and Tully, J.G. (eds): in The Mycoplasmas, Vol. V, Academic Press, New York, pp. 244–364.

17. Bowyer, J.W. and Calavan, E.C., 1974, Antibiotic sensitivity in vitro of the mycoplasma-like organism associated with citrus stubborn disease, *Phytopathology* **64**:346–349.

18. Brinton, L.P. and Burgdorfer, W., 1976, Cellular and subcellular organization of the 277 F agent: a spiroplasma from the rabbit tick, *Haemaphysalis leporispalustris* (Acari: lxodidae), *Int. J. Syst. Bacteriol.* **26**:554–560.

19. Burges, H.D., 1982, Control of insects by bacteria, *Parasitology* **84**:79–117.

20. Chang, C.J. and Chen, T.A., 1978, Antibiotic spectrum of pathogenic spiroplasmas of plants, insect and vertebrate animals, *Phytopathology News.* **12**:233.

21. Chastel, C., Gilot, B., Le Goff, F., Gruffaz, R., and Abalain-Colloc, M.L., 1985, Isolement de spiroplasmes en France (Savoie, Alpes du Nord) à partir de moustiques du genre *Aedes, C.R. Acad. Sci., Paris* **300**:261–266.

22. Chastel, C., Abalain-Colloc, M.L., Gilot, B., Le Goff, F., Gruffaz, R., and Simitzis, A.M., 1985, Spiroplasmes et arthropodes hématophages: Perspectives en pathologie tropicale, *Bull. Soc. Pathol. Exot.* **78**:769–779.

23. Chastel, C., Le Goff, F., and Humphery-Smith, I., Multiplication and persistence of *Spiroplasma melliferum*, strain A 56, in experimentally infected suckling mice. 8th Intern. Cong. of the International Organization for Mycoplasmology, Istanbul, Turkey, July 8–12, 1990.

24. Chastel, C., Devau, B., Le Goff, F., Simitzis-Le Flohic, A.M., Gruffaz, R., Kerdraon, G., and Gilot, B., 1987, Mosquito spiroplasmas from France and their ecology, *Israel J. Med. Sci.* **23**:683–686.

25. Chastel, C., Gilot, B., Le Goff, F., Devau, B., Kerdraon, G., Humphery-Smith, I., and Simitzis-Le Flohic, A.M., 1990, New developments in the ecology of mosquito spiroplasmas, *Zbl. Bakt. Mikrobiol. Hyg. Abt. I. Suppl* *20*:455–460.

26. Chastel, C., Le Goff, F., Goas, J.Y., Goasgen, J., Kerdraon, G., and Simitzis-Le Flohic, A.M., 1990, Serosurvey on the possible role of mosquito spiroplasmas in central nervous systems disorders in man, *Zbl. Bakt. Mikrob. Hyg. I. Abt. Suppl. 20*:910–911.

27. Chen, T.A. and Liao, C.H., 1975, Corn stunt spiroplasma; isolation, cultivation, and proof of pathogenicity, *Science* **188**:1015–1017.

28. Christiansen, C., Askaa, G., Freundt, E.A., and Whitcomb, R.F., 1979, Nucleic acid hybridization experiments with *Spiroplasma citri* and the corn stunt and suckling mouse cataract spiroplasmas, *Curr. Microbiol.* **2**:323–326.

29. Clark, H.F., 1964, Suckling mouse cataract agent, *J. Infect. Dis.* **114**:476–487.

30. Clark, H.F., 1969, Rat cataract induced by suckling mouse cataract agent, *Am. J. Ophtalmol.* **68**:304–308.

31. Clark, T.B., Peterson, B.V., Whitcomb, R.F., Henegar, R.B., Hackett, K.J., and Tully, J.G., 1984, Spiroplasmas in the Tabanidae, *Israel J. Med. Sci.* **20**:1002–1005.

32. Clark, T.B., Whitcomb, R.F., Tully, J.G., Mouches, C., Saillard, C., Bové, J.M., Wroblewski, H., Carle, P., Rose, D.L., Henegar, R.B., and Williamson, D.L., 1985, *Spiroplasma melliferum*, a new species from the honeybee (*Apis mellifera*), *Int. J. Syst. Bacteriol.* **35**:296–308.

33. Clark, T.B., Henegar, R.B., Rosen, L., Hackett, K.J., Whitcomb, R.F., Lowry, J.E., Saillard, C., Bové, J.M., Tully, J.G., and Williamson, D.L., 1987, New spiroplasmas from insects and flowers: isolation, ecology, and host association. *Israel J. Med. Sci.*, **23**:687–690.

34. Cole, R.M., Tully, J.G., Popkin, T.J., and Bové, J.M., 1973, Ultrastructure of the agent of citrus "stubborn" disease, *Ann. N.Y. Acad. Sci.* **225**:471–493.

35. Daniels, M.J., 1979, Mechanisms of spiroplasma pathogenicity, Whitcomb, R.F. and Tully, J.G. (eds): in The Mycoplasmas, Vol. III, Academic Press, New York, pp. 37–64.

36. Daniels, M.J., 1979, Mechanisms of spiroplasma pathogenicity, *J. Gen. Microbiol.* **114**:323–328.

37. Daniels, M.J., 1983, Mechanisms of spiroplasma pathogenicity, *Annu. Rev. Phytopathol.* **21**:29–43.

38. Davis, R.E., Worley, J.F., Whitcomb, R.F., Ishijima, T., and Steere, R.L., 1972, Helical filaments produced by a mycoplasma-like organism associated with corn stunt disease, *Science* **176**:521–523.

39. Davis, R.E. and Worley, J.F., 1973, Spiroplasma: motile helical microorganism associated with corn stunt disease, *Phytopathology* **63**:403–408.

40. Davis, R.E., 1981, Antibiotic sensitivities *in vitro* of diverse spiroplasma strains associated with plants and insects, *Appl. Environ. Microbiol.* **41**:329–333.

41. Eskafi, F.M., Mc Coy, R.E., and Norris, R.C., 1987, Pathology of *Spiroplasma floricola* in *Galleria mellonella* larvae, *J. Invert. Pathol.* **49**:1–13.

42. Fields, B.N., 1987, Powerful prions? *N. Engl. J. Med.* **317**:1597–1598.

43. Fitz-Earle, M., Sakaguchi, B., Horio, M., and Toukamoto, M., 1987, Transfer of *Drosophila* (Diptera: Drosophilidae) Sex Ratio spiroplasmas into *Toxorhynchites splendens* (Diptera: Culicidae). Evidence for transovarial transmission but absence of sex ratio distortion, *J. Med. Entomol.* **24**:448–457.

44. Efrench-Constant, R.H. and Devonshire, A.L., 1987, A multiple homogeniz-

er for the rapid preparation of samples for immunoassays and electrophoresis, *Biochem. Genet.* **25**:493–499.

45. Freund, E.A., Whitcomb, R.F., Barile, M.F., Razin, S., and Tully, J.G., 1984, Proposal for elevation of the family *Acholeplasmataceae* to ordinal rank: *Acholeplasmatales, Int. J. Syst. Bacteriol.* **34**:346–349.

46. Friedlaender, R.P., Barile, M.F., Kuwabara, T., and Clark, H.F., 1976, Ocular pathology induced by the suckling mouse cataract agent, *Invest. Ophthalmol.* **15**:640–647.

47. Fudl-Allah, A.A., Calavan, E.C., and Igwegbe, E.C.K., 1972, Culture of a mycoplasm-like organism associated with stubborn disease of citrus, *Phytopathology* **62**:729–731.

48. Gajdusek, D.C., 1977, Unconventional viruses and the origin and disappearance of Kuru, *Science* **197**:943–960.

49. Gajdusek, D.C., 1985, Unconventional viruses causing subacute spongiform encephalopathies, Fields, B.N., Knipe, D.M., Melnick, J.L., Chanock, R.M., Roiznon, B., and Shope, R.E. (eds): in Virology, Raven Press, New York, pp. 1519–1557.

50. Garnier, M., Bébéar, C., Latrille, J., and Bové, J.M., 1978, Morphologie et croissance des spiroplasmes, *Bordeaux Méd.* **11**:1781–1786.

51. Garnier, M., Clerc, M., and Bové, J.M., 1981, Growth and division of spiroplasmas: Morphology of *Spiroplasma citri* during growth in liquid medium, *J. Bacteriol.* **147**:642–652.

52. Garnier, M., Clerc, M., and Bové, J.M., 1984, Growth and division of spiroplasmas: elongation of elementary helices, *J. Bacteriol.* **158**:23–28.

53. Garnier, M., Steiner, T., Martin, G., and Bové, J.M., 1984, Oxidoreduction with insect cells in culture, *Israel J. Med. Sci.* **20**:840–842.

54. Giannotti, J. and Giannotti D., 1986, Multiplication de mollicutes *in situ* et *in vitro* dans le nectar floral de différentes plantes, *C.R. Acad. Sciences, Paris* **302**:669–674.

55. Gray, A., Francis, R.J., and Scholtz, C.L., 1980, Spiroplasma and Creutzfeldt-Jakob disease, *Lancet* **ii**:152.

56. Greig, R.G., Koestler, T.P., Trainer, D.L., Carwin, S.P., Miles, L., Kline, T., Sweet, R., Yokoyuma, S., and Poste, G., 1985, Tumorigenic and metastatic properties of normal and *ras*-transfected NIH/3T3 cells, *Proc. Nat. Acad. Sci., USA* **82**:3698–3701.

57. Grimstad, P.R. and De Foliart, G.R., 1974, Nectar sources of Wisconsin mosquitoes, *J. Med. Entomol.* **11**:331–341.

58. Grulet, O., Humphery-Smith, I., and Chastel, C., A spiroplasma specific medium. 8th Intern. Cong. of the International Organisation for Mycoplasmology, Istanbul, Turkey, July 8–12, 1990.

59. Gubler, D.J., 1989, *Aedes aegypti* and *Aedes aegypti*-borne disease control in the 1990's top down or bottom up, *Am. J. Trop. Med. Hyg.* **40**:571–579.

60. Hackett, K.J. and Lynn, D.E., 1985, Cell-assisted growth of a fastidious spiroplasma, *Science* **230**:825–827.

61. Hackett, K.J., Lynn, D.E., Williamson, D.L., Ginsberg, A.S., and Whitcomb, R.F., 1986, Cultivation of the Drosophila sex-ratio spiroplasma, *Science* **232**:1253–1255.

62. Hackett, K.J., Ginsberg, A.S., Rottem, S., Henegar, R.B., and Whitcomb, R.F., 1987, A defined medium for a fastidious spiroplasma, *Science* **237**:525–527.

63. Hackett, K.J. and Clark, T.B., 1989, Ecology of spiroplasma, Whitcomb, R.F. and Tully, J.G. (eds): in The Mycoplasma, Vol. V, Academic Press, New York, pp. 113–200.

64. Halstead, S.B., 1987, Dengue hemorrhagic fever: Why can't we control it? St George, T.D., Kay, B.H. and Blok, J. (eds): in Arbovirus research in Australia. Commonwealth Scientific and Industrial Research Organisation and Queensland Institute of Medical Research, Brisbane, pp. 30–35.

65. Halstead, S.B., 1988, Pathogenisis of Dengue: Challenges of molecular biology, *Science* **239**:476–481.

66. Hardy, J.L., Houk, E.J., Kramer, L.D., and Reeves, W.C., 1983, Intrinsic factors affecting vector competence of mosquitoes for arboviruses, *Annu. Rev. Entomol.* **28**:229–262.

67. Helmuth, R., Stephan, R., Bunge, C., Hoog, B., Steinbeck, A., and Bulling, E., 1985, Epidemiology of virulence-associated plasmids and outer membrane protein patterns within seven common *Salmonella* serotypes, *Infect. Immun.* **48**:175–182.

68. Holt, T.A. and Phillips, J., 1988, Bovine spongiform encephalopathy, *Brit. Med. J.* **296**:1581–1582.

69. Humphery-Smith, I. and Chastel, C., 1988, Creutzfeldt-Jakob disease, spiroplasmas, and crystalline artifacts, *Lancet* November 19:1199.

70. Humphery-Smith, I., Grulet, O., Le Lay, G., and Chastel, C., 1988, Pathogenicité de *Spiroplasma sabaudiense* (Mollicutes) pour les cellules (C6/36) d'*Aedes albopictus* (Insecta: Diptera) *in vitro*, *Bull. Soc. Pathol. Exot.* **81**:752–765.

71. Humphery-Smith, I., Chastel, C., Grulet, O., and Le Goff, F., 1990, The infective cycle of *Spiroplasma sabaudiense in vitro* in *Aedes albopictus* (C6/36) cells and its potential for the biological control of Dengue virus, *Zentralbl. Bakteriol. Mikrobiol. Hyg. Abt. I. Suppl.*:922–924.

73. Humphery-Smith, I., Grulet, O., Le Goff, F., and Chastel, C., 1990, Spiroplasma (Mollicutes: Spiroplasmataceae) pathogenic for *Aedes aegypti* and *Anopheles stephensi* (Diptera: Culicidae). *J. Med. Entomol.*

74. Humphery-Smith, I., Grulet, O., and Chastel, C., Persistence of a mosquito spiroplasma (Class:Mollicutes). *Appl. Environ. Microbiol.*

75. Humphery-Smith, I., Grulet, O., and Chastel, C., With press, Pathogenicity of *Spiroplasma taiwanense* (Class: Mollicutes) for larval *Aedes aegypti* (Diptera: Culicidae), *J. Invert. Pathol.*

76. Humphery-Smith, I., Grulet, O., Le Goff, F., and Chastel, C., In Prep., Pleomorphism and persistence of *Spiroplasma sabaudiense* (Class: Mollicutes). *Ann. Microbiol.*

77. Humphery-Smith, I., Le Goff, F., Leclercq, M., and Chastel, C., Spiroplasmas from European Tabanidae, *Med. Vet. Entomol.*

78. Hung, S.H.Y., Chen, T.A., Whitcomb, R.F., Tully, J.G., and Chen Y.X., 1987, *Spiroplasma culicicola* sp. nov. from the salt marsh mosquito *Aedes sollicitans*, *Int. J. Syst. Bacteriol.* **37**:365–370.

79. International Committee on Systematic Bacteriology Subcommittee on the Taxonomy of *Mollicutes*, 1979, Proposal of minimal standards for descriptions of new species of the class *Mollicutes*, *Int. J. Syst. Bacteriol.* **29**:172–180.

80. Itoh, K., Pan, I.J., and Koshimizu, K., 1989, A proposed life cycle model of *Spiroplasma mirum* based on scanning electron microscopical observations of growth in liquid culture, *Microbiol. Immunol.* **33**:821–832.

81. Jones, G.W., Rabert, D.K., Svinarich, D.M., and Whitfield, H.J., 1982, Association of adhesive, invasive and virulent phenotypes of *Salmonella typhimurium* with autonomous 60-megadalton plasmids, *Infect. Immun.* **38**:476–486.

82. Kirchhoff, H., Heitmann, J. and Trautwein, G., 1981, Pathogenicity of *Spiroplasma* sp. strain SMCA in rabbits: clinical, microbiological, and histological aspects, *Infect. Immun.* **33**:292–296.

83. Kirchhoff, H., Kuwabara, T., and Barile, M.F., 1981, Pathogenicity of *Spiroplasma* sp. strain SMCA in Syrian hamsters: clinical, microbiological and histological aspects, *Infect. Immun.* **31**:445–452.

84. Kondo, F. and Maramorosch, K., 1977, Antibiotic spectrum of plant pathogenic spiroplasmas, *Ann. Soc. Microbiol., Ann. Meeting* Abstract **G 19**:133.

85. Kotani, H., Phillips, D., and Mc Garrity, G.J., 1986, Malignant transformation of NIH-3T3 and CV-1 cells by a helical mycoplasma, *Spiroplasma mirum*, strain SMCA, *In Vitro Cell. Dev. Biol.* **22**:756–762.

86. Kouri, G.P., Guzman, M.G., Bravo, J.R., and Triana, C., 1989, Dengue haemorrhagic fever/dengue shock syndrome: lessons from the Cuban epidemic, 1981, *Bull. W.H.O.* **67**:375–380.

87. Labarere, J. and Barroso, G., 1984, Ultraviolet irradiation mutagenesis and recombination in *Spiroplasma citri*, *Israel J. Med. Sci.* **20**:826–829.

88. Laflèche, D. and Bové, J.M., 1970, Mycoplasmes dans les agrumes atteints de "Greening," de "Stubborn" ou de maladies similaires, *Fruits* **25**:455–465.

89. Leach, R.H., Matthews, W.B., and Will, R., 1983, Creutzfeldt-Jakob disease: Failure to detect spiroplasmas by cultivation and serological tests, *J. Neurol. Sci.* **59**:349–353.

90. Le Goff, F., Marjolet, M., Guilloteau, J., Humphery-Smith, I., and Chastel, C., 1990, Characterization and ecology of mosquito spiroplasmas from Atlantic biotops in France. *Ann. Parasitol. Hum. Comp., Paris.*

91. Liao, C.H. and Chen, T.A., 1977, Culture of corn stunt spiroplasma in a simple medium. *Phytopathology*, **67**:802–807.

92. Liao, C.H. and Chen, T.A., 1981, In vitro susceptibility and resistance of two spiroplasmas to antibiotics, *Phytopathology* **71**:442–445.

93. Mc Garrity, G.J. and Kotani, H., 1984, Use of cell cultures to study spiroplasma infections, *Israel J. Med. Sci.* **20**:924–926.

94. Mc Garrity, G.J., Megraud, F., and Gamon, L., 1984, Rabbit lens cultures in the characterizarion of *Spiroplasma mirum* pathogenicity, *Annu. Microbiol.* **135A**:249–254.

95. Mc Garrity, G.J. and Kotani, H., 1985, Cell culture mycoplasmas, Razin, S. and Barile, M.F., (eds): in The Mycoplasmas, Vol. IV. Academic Press, New York, pp. 353–390.

96. Mc Garrity, G.J., Gamon, L., Steiner, T., Tully, J.G., and Kotani, H., 1985, Uridine phosphorylase activity among the class Mollicutes, *Curr. Microbiol.* **12**:107–112.

97. Magnarelli, L.A., 1977, Nectar feeding by *Aedes sollicitans* and its relation to gonotrophic activity, *Environ. Entomol.* **6**:237–242.

98. Magnarelli, L.A., 1978, Nectar feeding by female mosquitoes and its relation to follicular development and parity, *J. Med. Entomol.* **14**:527–530.

99. Mc Garrity, G.J. and Williamson, D.L., 1989, Spiroplasma pathogenicity *in*

vivo and *in vitro*, Whitcomb, R.F. and Tully, J.G. (eds): in The Mycoplasmas, Vol V, Academic Press, New York, pp. 365–392.

100. Manson-Bahr, P.E.C. and Bell, D.R., 1987, Manson's Tropical Diseases. London, Baillière Tindall, 1557p.

101. Megraud, F., Gamon, L.B., and Mc Garrity, G.J., 1983, Characterization of *Spiroplasma mirum* (Suckling Mouse Cataract Agent) in a rabbit lens cell culture, *Infect. Immun.* **42**:1168–1175.

102. Morgan, K.L., 1988, Bovine spongiform encephalopathy: time to take scrapie seriously, *Vet. Rec* **122**:445–446.

103. Mouches, C., Vignault, J.C., Tully, J.G., Whitcomb, R.F., and Bové, J.M., 1979, Characterization of spiroplasma by one- and two-dimensional protein analysis on polyacrylamide slab gels. *Curr. Microbiol.* **2**:69–74.

104. Narang, H.K., Asher, D.M., and Gajdusek, D.C., 1988, Evidence that DNA is present in abnormal tubulofilamentous structures found in scrapie, *Proc. Nat. Acad. Sci., USA* **85**:3575–3579.

105. Pardon, P., Popoff, M.Y., Coynault, C., Marly, J., and Miras, I., 1986, Virulence-associated plasmids of *Salmonella* serotype typhimurium in experimental murine infections, *Ann. Microbiol.* **137B**:47–60.

106. Pickens, E.G., Gerloff, R.K., and Burgdorfer, W., 1968, Spirochete from the rabbit tick, *Haemophysalis leporispalustris* (Packard), *J. Bacteriol.* **95**:291–299.

107. Pollack, J.D., Mc Elwain, M.C., De Santis, D., Manolukas, J.T., Tully, J.G., Chang, C.J., Whitcomb, R.F., Hackett, K.J., and Williams, M.V., 1989, Metabolism of members of the Spiroplasmataceae, *Int. J. Syst. Bacteriol.* **39**:406–412.

108. Prusiner, S.B., 1987, Prions and neurodegenerative diseases, *N. Engl. J. Med.* **317**:1571–1581.

109. Race, R.E., Caughey, B., Graham, K., Ernst, D., and Chesebro, B., 1988, Analysis of frequency of infection, specific infectivity, and prion protein biosynthesis in scrapie-infected neuroblastoma cell clones, *J. Virol.* **62**:2845–2849.

110. Razin, S. and Freundt, E.A., 1984, The mycoplasmas, Krieg, N.R. (ed): in Bergey's Manual of Systematic Bacteriology. Vol. 1, Williams and Wilkins, Baltimore/London, pp. 740–793.

111. Razin, S., 1985, Mycoplasma adherence, Razin, S., and Barile, M.F., (eds): in The Mycoplasmas, Vol. IV, Academic Press, New York, pp. 161–202.

112. Razin, S., 1988, Cell biology of Mycoplasmas: a synopsis. *Antibiot. Monitor*, **IV**:42–43.

113. Razin, S., 1989, Molecular approach to mycoplasma phylogeny, Whitcomb, R.F. and Tully, J.G., (eds): in The Mycoplasmas, Vol. V, Academic Press, New York, pp. 33–69.

114. Reeves, W.C., 1989, Concerns about the future of medical entomology in tropical medicine research, *Am. J. Trop. Med. Hyg.* **40**:569–570.

115. Reyes, J.M. and Hoenig, E.M., 1981, Intracellular spiral inclusions in cerebral cell processes in Creutzfeldt-Jakob disease, *J. Neuropath. Exp. Neurol.* **40**:1–8.

116. Rogers, M.J., Simmons, J., Walker, R.T., Weisburg, W.G., Woese, C.R., Tanner, R.S., Robinson, I.M., Stahl, D.A., Olsen, G., Leach, R.H., and Maniloff, J., 1985, Construction of the mycoplasma evolutionary tree from 5S

rRNA sequence data, *Proc. Nat. Acad. Sci. U.S.A.*, **82**:1160–1164.

117. Rosen, L., 1988, Further observations on the mechanism of vertical transmission of flaviviruses by *Aedes* mosquitoes, *Am. J. Trop. Med. Hyg.*, **39**:123–126.

118. Sabin, A.B., 1952, Research on Dengue during World War II, *Am. J. Trop. Med. Hyg.* **1**:30–50.

119. Saglio, P., Laflèche, D., Bonnissol, C., and Bové, J.M., 1971, Isolement, culture et observation au microscope électronique des structures de type mycoplasme associées à la maladie du Stubborn des agrumes et leur comparaison avec les structures observées dans le cas de la maladie du greening des agrumes, *Physiol. Vég.* **9**:569–582.

120. Saglio, P., L'Hospital, M., Laflèche, D., Bové, J.M., Tully, J.G., and Freundt, E.A., 1973, *Spiroplasma citri* gen. and sp. n: a mycoplasma-like organism associated with "stubborn" disease of citrus. *Int. J. Syst. Bacteriol.* **23**:191–204.

121. Saillard, C., Dunez, J., Garcia-Jurado, O., Nhami, A., and Bové, J.M., 1978, Détection de *Spiroplasma citri* dans les agrumes et les pervenches par la technique immuno-enzymatique "ELISA". *C.R. Acad. Sci., Paris*, **286**:1245–1248.

122. Schatzmayr, H.G., Nogueira, R.M.R., and Travassos da Rosa, A.P.A., 1986, An outbreak of Dengue virus at Rio de Janeiro, *Mem. Inst. Oswaldo Cruz, Rio J.* **81**:245–246.

123. Schwartz, J. and Elizan, T.S., 1972, Further characterization of Suckling Mouse Cataract Agent (SMCA): a slow, persistent infection of the nervous system, *Proc. Soc. Exp. Biol. Med.* **141**:699–704.

124. Shaikh, A.A., Johnson, W.E., Stevens, C., and Tang, A.Y., 1987, Isolation of spiroplasmas from mosquitoes in Macon County, Alabama. *J. Am. Mosq. Control. Assoc.*, **3**:289–295.

125. Siler, J.F., Hall, M.W., and Hitchens, A.P., 1925, Results obtained in the transmission of Dengue fever, *JAMA* **84**:1163–1172.

126. Simitzis-Le Flohic, A.M., Devaux B., Prigent, Y., Gruffaz, R., Le Goff, F., Gilot, B., and Chastel, C., 1988, Etude de l'effet pathogène expérimental des spiroplasmes isolés à partir de moustiques sur l'éclosion d'œufs d'*Aedes aegypti* et le devenir des larves issues de ces œufs, *Ann. Parasitol. Hum. Comp., Paris* **63**:76–84.

127. Skripal, I.G., 1974, On improvement of taxonomy of the class Mollicutes and establishment in the order Mycoplasmatales of the new family Spiroplasmataceae Fam. Nova. *Mikrobiologii Zh., Kiev*, **36**:462–467.

128. Skripal, I.G., 1983, Revival of the name *Spiroplasmataceae* fam. nov., nom. rev. omitted from the 1980 approved lists of bacterial names. *Int. J. Syst. Bacteriol.* **33**:408.

129. Slaff, M. and Chen, T.A., 1982, The isolation of a spiroplasma from *Aedes sollicitans* (Walker) in New Jersey. *J. Florida Anti-mosq. Assoc.*, **53**:19–21.

130. Stanek, G., Hirschl, A., and Laber G., 1981, Sensitivity of various spiroplasma strains against ethanol, formalin, glutaraldehyde, and phenol, *Zentralbl Bakteriol Hyg., Abt. I. Orig* **174B**:348–354.

131. Steiner, T., Mc Garrity, G.J., and Phillips, D.M., 1982, Cultivation and partial characterization of spiroplasmas in cell cultures, *Infect. Immun.* **35**:296–304.

132. Steiner, T., Mc Garrity, G.J., Bové, J.M., Phillips, D.M., and Garnier, M., 1984, Insect cell cultures in the study of attachment and pathogenicity of spiroplasmas and mycoplasmas, *Ann. Microbiol.* **135A**:47–53.
133. Tully, J.G., Whitcomb, R.F., Williamson, D.L., and Clark, H.F., 1976, Suckling mouse cataract agent is a helical wall-free prokaryote (Spiroplasma) pathogenic for vertebrates, *Nature* **259**:117–120.
134. Tully, J.G., Whitcomb, R.F., Clark, H.F., and Williamson, D.L., 1977, Pathogenic mycoplasmas: cultivation and vertebrate pathogenicity of a new spiroplasma, *Science* **195**:892–894.
135. Tully, J.G., Rose, D.L., Yunker, C.E., Cory, J., Whitcomb, R.F., and Williamson, D.L., 1981, Helical mycoplasmas (Spiroplasmas) from *Ixodes* ticks, *Science* **212**:1043–1045.
136. Tully, J.G. and Whitcomb, R.F., 1981, The genus *Spiroplasmas*, Starr, M.P., Stolp, H., Trüper, H.G., Balows, A. and Schelgel, H.G., (eds): in The Prokaryotes: A Handbook on Habitats, Isolation and Identification of Bacteria. Springer–Verlag, Berlin, pp. 2271–2284.
137. Tully, J.G., Whitcomb, R.F., Rose, D.L., and Bové, J.M., 1982, *Spiroplasma mirum*, a new species from the rabbit tick (*Haemaphysalis leporispalustris*), *Int. J. Syst. Bacteriol.* **32**:92–100.
138. Tully, J.G., Rose, D.L., Clark, E., Carle, P., Bové, J.M., Henegar, B., Whitcomb, R.F., Colflesh, D.E., and Williamson, D.L., 1987, Revised group classification of the genus *Spiroplasma* (Class *Mollicutes*), with proposed new groups XII to XXIII. *Int. J. Syst. Bacteriol.* **37**:357–364.
139. Tully, J.G., 1989, Class Mollicutes: new perspectives from plant and arthropod studies, Whitcomb, R.F. and Tully, J.G. (eds): in The Mycoplasmas, Vol V., Academic Press., New York, pp. 1–31.
140. Van Handel, E., 1972, The detection of nectar in mosquitoes, *Mosq. News* **32**:458.
141. Vargo, A.M. and Foster, W.A., 1984, Gonotrophic state and parity of nectar-feeding mosquitoes, *Mosq. News* **44**:6–10.
142. Wallace, D.C. and Morowitz, H.J., 1973, Genome size and evolution. *Chromosoma*, **40**:121–126.
143. Wells, G.A.H., Scott, A.C., Johnson, C.T., Gunning, R.F., Hancock, R.D., Jeffrey, M., Dawson, M., and Bradley, R., 1987, A novel progressive spongiform encephalopathy in cattle, *Vet. Rec.* **121**:415–420.
144. Whitcomb, R.F., Tully, J.G., Mc Cawley, P., and Rose, D.L., 1982, Application of the growth inhibition test to *Spiroplasma* taxonomy, *Int. J. Syst. Bacteriol.* **32**:387–394.
145. Whitcomb, R.F., Clark, T.B., Tully, J.G., Chen, T.A., and Bové, J.M., 1983, Serological classification of spiroplasmas: current status. *Yale. J. Biol. Med.*, **56**:453–459.
146. Whitcomb, R.F., Chen, T.A., Williamson, D.L., Liao, C., Tully, J.G., Bové, J.M., Mouches, C., Rose, D.L. Coan, M.E., and Clark, T.B., 1986, *Spiroplasma kunkelii* sp. nov., the etiological agent of corn stunt disease, *Int. J. Syst. Bacteriol.* **36**:170–178.
147. Whitmore, S.C., Rissler, J.F., and Davis, R.E., 1983, In vitro susceptibility of spiroplasmas to heavy-metal salts, *Antimicrob. Agents. Chemother.* **23**:22–25.
148. Williamson, D.L. and Whitcomb, R.F., 1975, Plant mycoplasmas: Cultivable

spiroplasma causes corn stunt disease, *Science* **188**:1018–1020.

149. Williamson, D.L., Whitcomb, R.F., and Tully, J.G., 1978, The spiroplasma deformation test, a new serological method, *Curr. Microbiol.* **1**:203–207.

150. Williamson, D.L. and Poulson, D.F., 1979, Sex Ratio Organisms (Spiroplasmas) of Drosophila, Whitcomb, R.F., and Tully, J.G., (eds): in The Mycoplasmas, Vol. III, Academic Press, New York, pp. 175–208.

151. Williamson, D.L., Tully, J.G., and Whitcomb, R.F., 1989, The genus *Spiroplasmas*, Whitcomb, R.F. and Tully, J.G., (eds): in The Mycoplasmas, Vol. V, Academic Press, New York, pp. 71–111.

152. Woese, C.R., Maniloff, J., and Zablen, L.B., 1980, Phylogenetic analysis of the mycoplasmas, *Proc. Natl. Acad. Sci., U.S.A.*, **76**:381–385.

153. Woese, C.R., Stackebrandt, E., and Ludwig, W., 1985, What are mycoplasmas: the relationship of tempo and mode in bacterial evolution, *J. Mol. Evol.*, **21**:305–316.

154. Woese, C.R., 1987, Bacterial evolution, *Microbiol. Rev.*, **51**:221–271.

155. Wroblewski, H., Robic, D., Thomas, D., and Blanchard A., 1984, Comparison of the amino-acid compositions and antigenic properties of spiralins purified from the plasma membranes of different spiroplasmas, *Ann. Microbiol.* **135A**:73–82.

156. Yunker, C.E., Tully, J.G., and Cory, J., 1987, Arthropod cell lines in the isolation and propagation of tick-borne spiroplasmas, *Curr. Microbiol.* **15**:45–50.

Index

Acholeplasmataceae, 153
Acholeplasmatales, 153
Actual isolation rate (AIR), 159
Aedes aegypti, 104, 195
Aedes sollicitans, 151
African trypanosomiasis, 117–139
AIR (actual isolation rate), 159
Amino acids, 113
Anaeroplasmataceae, 153
Anaeroplasmatales, 153
Anopheles albimanus, 104
Anoplura, 103
Aphid-borne potato viruses, 53–68
Aphid effects, intrinsic, 34–36
Aphid population dynamics
  biotic factors and, 34–38
  crop management and, 38–42
  virus spread and, 19–43
  weather and, 22–34
Aphids, 19–20
  natural enemies of, 36–37
Arthropod cell lines, 162–164
Artificial diets for blood-feeding
  insects, 103–113

Bacillus thuringiensis, 195–196
Barley yellow dwarf virus (BYDV),
  vii, 21
  effect on winter cereal survival,
  77–87
  epidemiology, 75–77

metabolic changes in cereals due
  to, 87–95
tolerance, 79–81
winter-stress tolerances and,
  73–96
Beet mild yellowing virus (BMYV),
  22
Beet yellows virus (BYV), 22
Beetle-transmitted viruses, 15
Blood composition, 106
Blood dietary studies
  approaches for, 105
  classification of, 104–105
Blood feeding, 103
Blood-feeding insects, artificial
  diets for, 103–113
BMYV (beet mild yellowing virus),
  22
Bovine blood, 106–109
Bovine erythrocytes, 107
British sugarbeet yellows cycle, 65
BYDV, *see* Barley yellow dwarf
  virus
BYV (beet yellows virus), 22

Cabbage mosaic virus (CMV), 32
Canada wildlife cycle, 63–64
Cantharis spiroplasma, 156, 184
Cereals, metabolic changes due to
  BYDV in, 87–95
Cholesterol, 112